我国土著鱼类洛氏鱥的低氧耐受生理及其分子机制研究

母伟杰 著

中国农业出版社
农村读物出版社
北 京

前　言

洛氏鱥为小型冷水性鱼类，自然分布在黑龙江、图们江、辽河、黄河以及长江中上游的支流，属于黑龙江省较有特色的温水冷水性鱼类。由于高纬度广泛分布，且耐受低温、低氧能力较强，其具有较好的高纬度水域耐受性分析价值。此外，洛氏鱥具有适应范围广、抗病力强、耐低氧等特点，是池塘、网箱及工业化养鱼的主要对象，适合集约化养殖，是极具养殖发展前景的鱼类品种。由于纬度较高，黑龙江省水生生态系统昼夜温差和含氧量变化剧烈，呈现早低、中午高、夜间最低的状况，部分地方性鱼类栖息在这些独特的缺氧环境中。分析鱼类适应缺氧的机制将更好地了解缺氧胁迫下的营养需求和利用，以及掌握鱼类缺氧适应策略的演变。但是，关于冷水鱼类耐受低氧的相关研究比较缺乏，尤其是从组织学、生理学和分子生物学等方面开展的综合分析较为缺乏。

基于上述研究现状，本书梳理了团队近年来开展的关于洛氏鱥低氧耐受方面的研究，主要从低氧应激、长期持续低氧和昼夜循环低氧方面展开。第一章和第二章介绍了洛氏鱥的生物学特性、养殖现状及养殖前景。第三章介绍了洛氏鱥经短期低氧胁迫后的组织、生理和分子应答。第四章和第五章介绍了持续低氧和昼夜循环低氧处理下，洛氏鱥鳃和肝脏、心脏和脑组织的组织、生理和分子应答。第六章介绍本团队近年来针对洛氏鱥低氧耐受性挖掘的基因，并分析了其组织表达模式和低氧胁迫后的表达趋势。

本专著承蒙国家自然科学基金面上项目《高纬度水域洛氏鱥耐受低氧环境的生理特征及分子响应机制》（32170523）资助。本书在撰写和修改过程中，得到了哈尔滨师范大学鱼类生理和免疫实验室研究生的大力支持，他们是杨宇婷、王振、王婧、姚铁辉、张天序和王思涵等。本书旨在展示高纬度鱼类洛氏鱥低氧耐受的生理策略等，其研究结果将为我国高纬度地区鱼类资源的开发和利用提供重要资料，并为全球范围内鱼类的适应进化研究提供参考。

书中所有素材均来自作者及所在团队的第一手研究成果，相关项目工作仍在进行中，新的调查结果将会在新成果中进行展示。由于作者的水平有限，本书不足之处敬请广大读者批评指正。

著　者

2022 年 4 月

目 录

前言

第一章　洛氏鱼岁概述 ··· 1
　一、洛氏鱼岁的主要生物学特征 ··· 1
　二、洛氏鱼岁的野外分布和种群资源特征 ································· 1
　三、洛氏鱼岁低氧耐受研究现状 ··· 2

第二章　洛氏鱼岁养殖研究及其发展前景 ··· 3
　第一节　洛氏鱼岁养殖潜力研究现状 ··· 3
　第二节　洛氏鱼岁饵料饲养和病害防治研究 ································· 4
　第三节　洛氏鱼岁产业发展现状及存在问题 ································· 4
　　一、洛氏鱼岁产业发展现状 ··· 4
　　二、洛氏鱼岁产业发展存在的问题 ··· 4

第三章　洛氏鱼岁耐受短期低氧胁迫的生理机制 ································· 6
　第一节　低氧处理后洛氏鱼岁的 LOE 分析和血液生理生化的影响 ······ 6
　　一、低氧处理后洛氏鱼岁 LOE 值 ··· 6
　　二、低氧处理对洛氏鱼岁血液生理生化的影响 ··························· 7
　第二节　低氧处理后鳃和肝脏组织形态学观察 ··························· 9
　　一、低氧处理后洛氏鱼岁鳃组织形态学观察 ······························· 9
　　二、低氧处理后洛氏鱼岁肝组织形态学观察 ····························· 12
　第三节　低氧胁迫对洛氏鱼岁肝脏糖脂代谢能力的影响 ················ 13
　第四节　低氧对洛氏鱼岁肝脏和鳃抗氧化能力和非特异性免疫能力的影响 ··· 17
　第五节　低氧处理后洛氏鱼岁肝脏中脂质代谢相关基因表达的变化 ······ 20

第四章　洛氏鱼岁持续低氧和昼夜低氧处理后鳃和肝脏生理应答机制比较 ··· 23
　第一节　持续低氧和昼夜低氧处理后 MO_2 值和血液学参数的变化 ··· 24
　第二节　持续低氧和昼夜低氧处理后洛氏鱼岁组织形态学观察 ········ 28

· 1 ·

一、持续低氧和昼夜低氧处理后洛氏鱥鳃组织形态学观察 ………… 28
　　二、持续低氧和昼夜低氧处理后洛氏鱥肝组织形态学观察 ………… 29
　　三、综合分析 ……………………………………………………………… 31
　第三节　持续低氧和昼夜低氧处理后肝脏糖脂代谢酶及抗氧化酶的活性
　　　　　变化 ……………………………………………………………… 33
　　一、持续低氧和昼夜低氧处理对洛氏鱥肝脏糖脂代谢能力的影响 …… 33
　　二、持续低氧和昼夜低氧处理对洛氏鱥肝脏抗氧化酶活性和脂质过氧化能
　　　　力的影响 ………………………………………………………………… 38
　第四节　持续低氧和昼夜低氧处理后肝脏糖代谢相关基因和钟基因表达
　　　　　的变化 …………………………………………………………… 38

第五章　洛氏鱥昼夜低氧和持续低氧处理后心脏和脑的响应机制 …… 45
　　一、昼夜低氧和持续低氧处理后心脏的组织学变化 ………………… 46
　　二、昼夜低氧和持续低氧处理后脑的组织学变化 …………………… 48
　　三、昼夜低氧和持续低氧处理后心脏和脑中抗氧化酶活性的变化 …… 51
　　四、昼夜低氧和持续低氧处理后脑和心脏中糖脂代谢酶指标的变化 … 54
　　五、昼夜低氧和持续低氧处理后脑和心脏中低氧信号传导基因表达的
　　　　变化 ……………………………………………………………… 56

第六章　洛氏鱥低氧耐受相关基因克隆及表达分析 ………………… 62
　第一节　缺氧诱导因子 HIF 基因克隆及表达分析 ……………………… 62
　　一、洛氏鱥 HIF-1α、HIF-2α 和 HIF-3α 基因全长克隆 ……………… 63
　　二、洛氏鱥 HIF-1α、HIF-2α 和 HIF-3α 基因的组织表达 …………… 66
　　三、洛氏鱥 HIF-1α、HIF-2α 和 HIF-3α 基因在不同低氧处理下的
　　　　表达模式 ………………………………………………………… 68
　第二节　P53 基因克隆及表达分析 ……………………………………… 72
　　一、洛氏鱥 P53 基因全长克隆 ………………………………………… 74
　　二、洛氏鱥 P53 基因组织表达结果 …………………………………… 77
　第三节　HO 基因的片段克隆及表达分析 ……………………………… 81
　　一、洛氏鱥 HO-1 和 HO-2 基因基本信息及同源性比较 …………… 82
　　二、洛氏鱥 HO-1 和 HO-2 基因的组织表达 ………………………… 84
　　三、洛氏鱥 HO-1 和 HO-2 基因在不同低氧处理下的表达模式 …… 85

参考文献 ………………………………………………………………… 89

第一章 洛氏鱥概述

一、洛氏鱥的主要生物学特征

洛氏鱥（*Phoxinus lagowskii*）［拉氏鱥（*Rhynchocypris lagowski*）是其同物异名］，俗称柳根，隶属于鲤形目（Cypriniformes）鲤科（Cyprinidae）雅罗鱼亚科（Leuciscinae）鱥属（*Phoxinus*）（周错，2018）。其身体长，稍侧扁，腹部圆，尾柄长而低。头近锥形，头长显著大于体高。吻尖，有时向前突出。口亚下位，口裂稍斜，上颌长于下颌。眼中等大，位于头侧的前上方。鳃孔中大，向前伸延至前鳃盖骨后缘的下方有膜与峡部相连。背鳍位于腹鳍的上方，外缘平直。臀鳍与背鳍同形，位于背鳍的后下方。胸鳍短，末端钝。鳃耙短小，排列稀。肠短，前后弯曲，其长短于体长，腹膜黑色。固定标本体背侧灰黑色，腹侧浅色，体侧常有疏散的黑色小点，背部正中自头后至尾鳍基有 1 黑色纵带，体侧自鳃孔上角至尾鳍基有 1 黑色纵带，尾部较为显著。尾鳍基部有 1 黑点，背鳍、尾鳍、胸鳍浅灰色，臀鳍、腹鳍浅色（赵文阁，2018）。洛氏鱥由于味道鲜美、肉质细嫩、营养价值高，富含蛋白质和不饱和脂肪酸，近年来在我国东北地区销量较好，是广大消费者喜食的小型名贵鱼类（瞿子惠，2019）。

洛氏鱥作为喜冷水的小型鱼类，在我国北方地区的繁殖一般始于 5 月初水温达到 12℃时（郭文学，2015），其个体繁殖力存在明显的区域性差异以及体长和体重依赖关系（张永泉，2015）。作为东北地区的名特优品种，一些学者研究了不同的饲料添加成分对洛氏鱥生理代谢、免疫和生长等方面的影响（陈秀梅，2018；杨兰，2018；瞿子惠，2019）。而作为各溪流水域的常见种或优势种，水温和流速可能是影响其生境适应的重要环境因子（康鑫，2011）（彩图 1）。

二、洛氏鱥的野外分布和种群资源特征

洛氏鱥主要分布在黑龙江、图们江、辽河、黄河以及长江中上游的支流（邹瑞兴等，2015），多栖息于水温较低，水质澄清，河底多页岩、砾石的小河，多石缝的山涧中或两河口交汇处（李月红等，2013）。洛氏鱥的最适生存温度为 10～25℃，喜集群生活在水流湍急且清澈的河流中。在天然水域中，洛氏鱥主要以水生昆虫幼虫为食，同时也食水生植物如蓝藻、绿藻、硅藻等，

是一种小型杂食性经济鱼类。

有研究认为，在最后一次冰川期间，洛氏鱥的迁徙路线的一部分被冰川封闭，从而隔离了该物种的一个亚群，使洛氏鱥没有连续分布在河流中，导致其在储层中长期隔离（Xue et al.，2017）。近年来，由于过度捕捞等行为，东北地区洛氏鱥的野生种群数量严重下降，一些种群已经完全消失。然而，有关该物种的信息仍然非常匮乏，迫切需要研究人员针对洛氏鱥的性状、生理机制、耐受环境能力和保护措施等开展研究。此外，鱥属鱼类大跨度的纬度分布，结合不断发展的分子生物技术以及生物信息分析，或许可以成为研究物种适应性进化的良好素材。

三、洛氏鱥低氧耐受研究现状

由于纬度较高，黑龙江省水生生态系统昼夜温差和含氧量变化剧烈，呈现早低、中午高、夜间最低的现象，部分地方性鱼类栖息在这些独特的缺氧环境中。洛氏鱥作为冷水洄游鱼类，其缺氧耐受性仍未得到研究。分析鱼类适应缺氧的机制将更好地了解缺氧胁迫下的营养需求和利用，以及掌握鱼类缺氧适应策略的演变。

第二章　洛氏鱥养殖研究及其发展前景

第一节　洛氏鱥养殖潜力研究现状

1960年，王岐山对我国分布的洛氏鱥进行了形态特征描述。查阅资料，国内外学者关于洛氏鱥的繁殖特征以及产业发展研究起步较晚，仅有我国学者从21世纪初期才开始对洛氏鱥开展繁殖方面的研究。早期的报道针对野生鱼类繁殖特征研究较多，如张永泉等（2015）研究报道了野生洛氏鱥的繁殖力及其与生物学指标的关系，发现在研究的流域洛氏鱥的个体绝对繁殖力主要分布在5 000～9 000粒，体重相对繁殖力主要分布在50～90粒/g，体长相对繁殖力主要分布在250～450粒/cm，个体绝对繁殖力和体长相对繁殖力均随体长、体重、成熟系数和肥满度的增加而极显著增加。王茂林等（2013）研究了本溪太子河流域洛氏鱥的个体繁殖力及其与生物学指标的关系，结果表明，洛氏鱥个体绝对繁殖力达到601.1～15 775.0粒，体长相对繁殖力（F_L）为75.8～1 132.9粒/cm，体重相对繁殖力（F_W）为64.7～602.3粒/g。此外，康鑫等对太子河洛氏鱥幼鱼栖息地适宜度进行了评估，结果表明，太子河幼鱼栖息地的适宜度指数，从上游到下游呈下降趋势，该流域内限制洛氏鱥幼鱼生存的因素主要是河流水质及河岸带等环境因素。张永泉等（2011）研究了温度和流速对洛氏鱥呼吸代谢的影响，结果表明水流流速12～16cm/s、温度16～24℃为1龄洛氏鱥理想养殖条件。

关于人工养殖的研究，杜业峰项目组（2019）经过多年研究，于2011年实现了洛氏鱥全人工繁殖和苗种培育。王春清等（2014）针对洛氏鱥的鱼池选择与建造、鱼的繁殖、鱼苗鱼种的放养及饲养管理等，以东北地区的池塘驯养方式进行了研究。张永泉等（2013）比较分析了雌雄洛氏鱥肌肉的营养成分，结果表明，洛氏鱥可作为优良的蛋白质和必需氨基酸的来源，雌性个体营养价值要优于雄性。郭文学等（2015）开展了黑龙江流域绥芬河水系野生洛氏鱥胚胎发育研究，结果发现，洛氏鱥成熟卵为圆球形，为黏性卵，卵膜适中，整个胚胎发育过程可分为受精卵、卵裂、囊胚、原肠胚、神经胚、器官形成、破膜等阶段。程湘军（2010）研究结果表明，洛氏鱥可以在池塘中进行主养和套养，能够驯化养殖，养殖成活率较高，养殖经济效益较好。

第二节　洛氏鱥饵料饲养和病害防治研究

近年来，关于洛氏鱥生理适应性的研究较为匮乏，而是集中在饵料饲养、病害防治以及地理分布等方面。例如，在评估大豆球蛋白对洛氏鱥幼鱼的生长、消化、免疫功能、抗氧化能力以及肝胰腺和肠道形态影响的研究中发现，日粮 40～80g/kg 甘氨酸显著降低鱼的增重率和比生长率，降低肝胰腺和肠道的蛋白酶活性和全身粗蛋白含量，降低过氧化氢酶、谷胱甘肽过氧化物酶和总超氧化物歧化酶活性，对于洛氏鱥幼鱼生长不利（Zhu et al.，2021）。通过膳食补充微生物絮凝物和铜暴露对洛氏鱥的炎症反应、氧化应激、肠道凋亡和屏障功能障碍的保护作用和潜在机制进行分析，结果发现，微生物絮凝物可提高溶菌酶、补体、免疫球蛋白和总抗氧化能力，且微生物絮凝物有助于调节 NF-κB/Nrf2 信号分子基因的表达（Yu et al.，2020）。评估了左旋肉碱对洛氏鱥生长和抗氧化功能的影响后，发现 750mg/kg 左旋肉碱使血清和肝胰腺中抗氧化酶活性显著高于对照组，且该左旋肉碱水平显著上调抗氧化酶和核因子的基因相对表达，从而得出结论，400～750mg/kg 的日粮左旋肉碱水平可以改善养殖条件下洛氏鱥的生长性能、饲料利用率和抗氧化防御系统（Wang et al.，2019）。洛氏鱥作为一种高纬度广泛分布的鱼类，且极具食用价值，分析其响应环境变化的策略具有一定的研究意义。

第三节　洛氏鱥产业发展现状及存在问题

一、洛氏鱥产业发展现状

洛氏鱥为冷水土著小型经济鱼类，自然状况下广泛分布于黑龙江流域（张永泉等，2013）。具有养殖适应范围广、起捕率高、抗病力强、耐低氧等特点，是池塘、网箱及工业化养鱼的主要对象，适合集约化养殖，是极具养殖发展前景的鱼类品种（杜业峰等，2019）。由于在自然水域主要栖息于河流中，所以在黑龙江省又被称为河柳（孔令杰等，2018）。洛氏鱥在黑龙江省有 30 年的养殖历史，2017 年被黑龙江省遴选为水产主导品种，为黑龙江名优特鱼类品种，在全省推广养殖，目前已经取得了显著成效。该品种因肉质细嫩、味道鲜美、肌间刺较少、蛋白质和不饱和脂肪酸含量高、口感佳而经济价值较高，黑龙江省商品鱼的出塘销售价格维持在 30 元/kg 左右（杜业峰等，2019；邹瑞兴等，2015）。

二、洛氏鱥产业发展存在的问题

近年来，随着生存环境污染、产卵场破坏和过渡捕捞等导致该鱼自然资源急剧减少（张永泉等，2013）。养殖区域分散不集中，养殖规模小、产量低等问题也制约着洛氏鱥的产业发展。例如，黑龙江省内的养殖户零星地分布在牡丹江、大兴安岭、绥化、大庆、哈尔滨、佳木斯和双鸭山等地，目前存在着养殖规模小、产量低、效益不高的特点（杜业峰等，2019）。此外，养殖技术水平较低和相关配套服务滞后也是制约发展的原因之一。例如，黑龙江省内专业化、规模化的柳根鱼苗种繁育和饲料生产企业以及病害防治的服务体系较为薄弱，针对洛氏鱥相关产品和技术服务支撑开展内容较少。在长期的饲养和研究过程中发现，洛氏鱥个体生长差异极其显著。目前养殖洛氏鱥的苗种来源主要是捕捞野生苗种，均未经过人工选育，使得养殖商品鱼大小参差不齐，影响了市场销售价格。因此，培育出生长速度快和生物学性状基本一致的洛氏鱥优良品种，已成为该鱼人工养殖亟须解决的瓶颈问题（张永泉等，2013）。

第三章 洛氏鲹耐受短期低氧胁迫的生理机制

近来，日益严重的缺氧现象被认为是全球水生态系统面临的最严重威胁之一（Diaz et al.，2009）。此外，营养物质过度排放到水生环境中导致的富营养化严重降低了河流中的溶解氧（Diaz and Breitburg，2009）。众所周知，温度升高也与氧溶解度降低有关，而这种生物需氧量的增加会导致更频繁和更严重的缺氧发作（Benson and Krause Jr，1984；Ficke et al.，2007）。缺氧可能会对鱼类栖息地质量和生物链产生负面影响，导致许多潜在的有害生理干扰，在极端情况下甚至会导致鱼类死亡（Pothoven et al.，2009；Roberts，2010；Sun et al.，2020）。明确鱼类生理适应性在缺氧耐受性方面的机制对于了解鱼类将如何应对日益普遍的缺氧环境非常重要（Collins et al.，2016；Vaquer-Sunyer and Duarte，2008）。

第一节 低氧处理后洛氏鲹的 LOE 分析和血液生理生化的影响

一、低氧处理后洛氏鲹 LOE 值

（一）低氧处理后洛氏鲹的 LOE 值

鱼体失去平衡的临界点（loss of equilibrium，LOE）是分析鱼类低氧耐受的指标之一。LOE 代表鱼类维持平衡的能力受到损害的氧气阈值（Dan et al.，2014）。彩图 2 显示了低氧胁迫 24h 后洛氏鲹的行为反应。鱼在溶解氧浓度>0.4mg/L 时，可以维持游泳行为。在溶解氧浓度为 0.7～0.8mg/L 时，鱼在水面上游动。在溶解氧浓度 0.6mg/L 时表现出一些剧烈游动。当氧气浓度为 0.4mg/L 时，洛氏鲹表现出明显的跳跃行为。最后，在氧分压为（0.21±0.01）mg/L 之前，洛氏鲹达到了 LOE 值。溶解氧浓度达到 0.28mg/L 后，第一条鱼在 57min 失去平衡，最后一条鱼在 60min 时失去平衡。

（二）综合分析

通常，LOE 被视为分析鱼类低氧耐受的指标之一（Wu et al.，2020）。在本研究中，洛氏鲹的 LOE 为 0.21mg/L，与许多鱼类物种相似或更低，如齐口裂腹鱼（*Schizothorax prenanti*）为 0.3～0.4mg/L（Fu et al.，2014），似

鳊驼背脂鲤（*Cyphocharax abramoides*）（1.7kPa，30℃）（Johannsson et al.，2018）和大菱鲆（*Scophthalmus maximus*）（1.5mg/L）（Jia et al.，2021），上述物种被认为是具有缺氧耐受性的物种。本研究中，洛氏鱥在缺氧条件下不会出现游泳速度缓慢等现象。根据文献，大多数低纬度和深海鱼类在缺氧条件下游泳活动减少以适应氧气供应量的减少，如尼罗罗非鱼（*Oreochromis niloticus*）（Li et al.，2018）和大西洋鳕（*Gadus morhua*）（Herbert and Steffensen，2005）等。然而，目前的研究发现，洛氏鱥在缺氧条件下移动非常剧烈，甚至表现出跳跃等行为。众所周知，缺氧会影响鱼类的行为并引发生理反应，而这种反应取决于特定物种的生理机能和适应能力（Chapman and McKenzie，2009；Roberts et al.，2012；Wannamaker and Rice，2000）。我们推测洛氏鱥强烈的行为反应与其栖息地水深有关，在栖息地洛氏鱥无法在水下进行大规模的垂直运动。激烈的运动和跳跃可以帮助洛氏鱥从长期的适应性进化中受益。之前的研究将鱼分成"大胆"和"害羞"的行为分类，研究中发现大胆的鱼更成功地获取资源（Laursen et al.，2011）。因此，我们可以认为洛氏鱥属于一种大胆（主动）的鱼，它可以通过行为改变在不断变化的环境中获得更多资源。

二、低氧处理对洛氏鱥血液生理生化的影响

鱼类采用复杂的生理调节策略，通过复杂的神经、行为和生理改变机制来应对缺氧应激，并试图维持体内平衡（Abdel-Tawwab et al.，2019）。鱼的血液学特征可以反映其生理状态和健康状况，因此被广泛用于识别或评估由不同压力因素引起的鱼类功能状态（Fazio，2019）。血细胞比容（Hct）、血红蛋白（HGB）、红细胞（RBC）和白细胞（WBC）等血液学参数用于评估鱼类携氧能力的功能状态。因此，血液学和生化参数与其他常规诊断方法相结合，可用于识别和评估导致缺氧应激和影响鱼类生产性能的条件。本研究中，通过测定洛氏鱥低氧胁迫后红细胞、白细胞、血细胞比容、血红蛋白、平均红细胞体积、红细胞平均血红蛋白浓度和平均红细胞血红蛋白含量等，探讨洛氏鱥低氧处理后的血液参数变化，分析洛氏鱥对低氧胁迫的耐受能力。

（一）低氧处理后洛氏鱥血液生理生化参数

低氧处理后，与对照相比，洛氏鱥红细胞显著降低，但再充氧组的RBC值恢复正常（表3-1）。与对照组相比，洛氏鱥处理组的白细胞WBC在24h 3mg/L低氧胁迫时增加，但是变化不显著。实验组的洛氏鱥平均红细胞体积（MCV）（0.5h 0.5mg/L）高于对照组。洛氏鱥血红蛋白在短期缺氧期间显著低于对照组，并在24h缺氧期间恢复到正常水平（$P<0.05$）。同时，洛氏鱥平均红细胞血红蛋白含量（MCH）在低氧处理期间

没有显著变化（$P<0.05$）。

表 3-1 低氧处理后洛氏鳄血液参数测定

指标	对照组	0.5h 3mg/L	0.5h 0.5mg/L	6h 3mg/L	24h 3mg/L	恢复组
RBC	2.89 ± 0.25^a	1.77 ± 0.36^b	1.80 ± 0.23^b	1.83 ± 0.2^b	2.38 ± 0.35^{ab}	2.89 ± 0.23^a
WBC	14.85 ± 1.33^{ab}	7.16 ± 1.4^a	13.12 ± 1.77^{ab}	11.25 ± 2.69^{ab}	18.43 ± 5.48^b	14.26 ± 4.1^{ab}
MCV	150.88 ± 5.91^a	171.67 ± 16.01^{ab}	187.2 ± 4.52^b	180.98 ± 4.93^{ab}	164.26 ± 11.61^{ab}	165.1 ± 9.82^{ab}
HGB	111.5 ± 2.31^b	81.33 ± 6.69^a	89.33 ± 5.81^a	88.6 ± 3.5^a	110 ± 5.57^b	113 ± 5.86^b
MCH	40.2 ± 1.44^a	47.97 ± 5.61^a	47.43 ± 4.24^a	51.18 ± 6.66^a	42.92 ± 6.5^a	42.24 ± 4.95^a

注：标有不同字母表示差异显著（$P<0.05$），否则差异不显著。

（二）综合分析

通常暴露于急性缺氧的鱼类通过调节红细胞等血液参数来维持对氧的吸收和向组织的输送氧气。缺氧时红细胞增多，可增强血液的携氧能力。研究表明，低氧胁迫后大西洋鲑（*Salmo salar*）（Wood et al., 2019）、团头鲂（*Megalobrama amblycephala*）（Wu et al., 2020）、中华鲟（*Acipenser schrenckii*）（Ni et al., 2019）等鱼类的红细胞、血红蛋白和红细胞比容比对照组高。红细胞的主要功能是转运氧气并介导呼吸中二氧化碳的产生（Cimen, 2008）。在红细胞中，血红蛋白充当氧气载体，通过结合氧气产生氧化的血红蛋白并随后在动物组织中释放氧气来发挥功能（Li et al., 2017）。

本研究结果表明洛氏鳄经低氧处理后，血红蛋白浓度在 24h 3mg/L 低氧胁迫时也有明显提高。白细胞是免疫系统循环细胞，参与先天性和获得性免疫反应，在免疫系统中发挥重要作用（Magadan, 2015）。缺氧应激增加了中华鲟（Ni et al., 2014）的白细胞数量，但在大菱鲆的研究中发现缺氧应激不影响大菱鲆的 WBC 数量。本研究中，经低氧处理后的洛氏鳄白细胞数量没有显著变化，这可能与低氧胁迫时间差异导致白细胞含量的不同有关，且反映物种特异性反应或物种耐受缺氧能力的差异。在团头鲂的研究中，与正常氧气条件处理下的红细胞数量和血红蛋白浓度相比，低氧处理后的红细胞数量和血红蛋白浓度显著增加（Wu et al., 2017）。然而，本研究结果显示，在 0.5h 3mg/L、0.5h 0.5mg/L 和 6h 3mg/L 组中，洛氏鳄的红细胞计数和血红蛋白浓度显著降低，而 0.5h 3mg/L 组白细胞计数显著下降，复氧组中白细胞计数恢复正常。缺氧-复氧会增加鱼血液中的红细胞比容、红细胞数量，这样的结果与许多研究一致（Mustafa et al., 2011；Riffel et al., 2014）。这些结果表明，缺氧-复氧通过增加鱼血中的红细胞数量和血红蛋白含量来增强氧气转运功能，缺氧-复氧对血液学指标影响的原因可能与鱼类的造血功能有关（Li et al., 2020）。在黑鲮脂鲤（*Prochilodus nigricans*）的研究中发现，红细胞的离散在低氧胁迫的 60min 达到最高水平，在正常氧气条件下恢复，表明急性低氧胁迫不会迅速诱导红细胞在缺氧条件下释放到循环中（Val et al.，2015）。在滨岸护胸鲶（*Hoplosternum littorale*）的研究中发现，低氧环境使鱼表现出明显的代谢抑制，从而减少对氧气的需求并减少红细胞生成（Brauner et al.，2011）。在本研究发现，平均红细胞体积略有增加，洛氏鲅可能通过减少红细胞的产生和增加红细胞的体积来应对低氧胁迫。脊椎动物细胞具有渗透干扰后立即激活的体积调节机制，细胞通过缓慢的体积下降对低渗肿胀作出反应，这被称为调节体积减少（regulatory volume decrease，RVD）反应（Cossins and Gibson，1997；Okada et al.，2001）。针对林蛙（*Rana temporaria*）的一项调查表明，在缺氧条件下，血红蛋白浓度没有发生显著变化，但细胞中调节体积的减少受到抑制，这可能与冬眠有关（Andreeva et al.，2018）。同样，冷水鱼洛氏鲅的血液学变化似乎代表了对其自然栖息地条件的基本适应。在长时间的低氧胁迫（72h）后，斑点叉尾鮰（*Ictalurus punetaus*）的平均红细胞血红蛋白含量和红细胞平均血红蛋白浓度显著增加，表明红细胞的携氧能力增加（Scott and Rogers，2006）。与大西洋鲑和隆头鱼（*Labrus bergylta*）的相关研究类似（Hvas et al.，2019），本研究结果表明洛氏鲅红细胞平均血红蛋白浓度低氧胁迫后有增加的趋势，但不显著，这可能与短期低氧胁迫有关。

第二节　低氧处理后鳃和肝脏组织形态学观察

本研究分析经低氧处理后，洛氏鲅中鳃的突出的薄片（PL）厚度、高度、基础长度，层间细胞团（ILCM）高度（μm），相邻薄片之间的距离（μm），ILCM 体积（μm^3），以及肝细胞和细胞核面积变化，进一步阐述洛氏鲅组织学响应低氧变化的机制。

一、低氧处理后洛氏鲅鳃组织形态学观察

（一）鳃组织形态学观察

低氧处理后洛氏鲅的鳃呈现明显的组织学改变（彩图3）。通过光镜及电镜观察，发现低氧处理后洛氏鲅鳃的 PL 厚度显著降低，PL 高度显著增加（图3-1、表3-2），恢复组表现出 PL 厚度增加和 PL 高度减少。突出薄片的基础长度在恢复组显著增加。与对照组相比，缺氧组的鳃中 ILCM 高度显著降低，在低氧处理组和恢复组中相邻薄片之间的距离显著增加（0.5h 3mg/L 除外）。并且，在低氧胁迫后也检测到突出薄片的面积增加和层间细胞团体积减少。在本研究中，在缺氧后计算鳃的气体扩散距离（GDD），所有缺氧鱼的 GDD 均低于对照组。

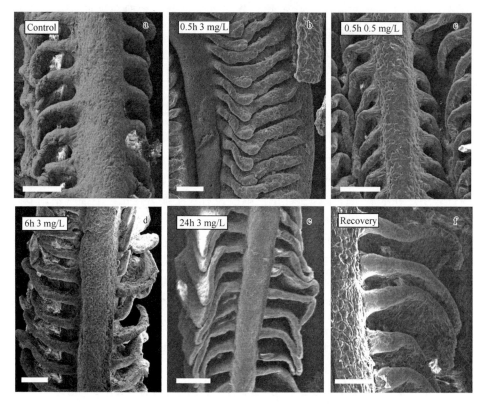

图 3-1　低氧暴露下洛氏鱥鳃扫描电镜（SEM）照片（标尺＝50μm）
a. 对照组　b. 0.5h 3mg/L 低氧处理　c. 0.5h 0.5mg/L 低氧处理
d. 6h 3mg/L 低氧处理　e. 24h 3mg/L 低氧处理　f. 恢复组

（二）综合分析

经短期低氧处理后，对洛氏鱥鳃的组织学分析，揭示了该鱼具有明显的低氧适应能力。结果表明，低氧胁迫显著增加了鳃的突出薄片的高度，并且减少了上皮肌丝的厚度，从而适应变化的氧气浓度。对鲤的研究结果证实，耐缺氧鲤的层状面积持续增加，表明鳃重塑是对缺氧的适应，通过增加呼吸表面积有助于其低氧的耐受性（Dhillon et al.，2013）。鲫和金鱼也会通过减少其层间细胞团的大小和增加鳃功能表面积来适应低氧（Sollid et al.，2003；Mitrovic et al.，2009）。这些策略用于增强从缺氧水中吸收氧或在低氧环境下维持常规代谢率的水平，从而延长生存时间（Nilsson，2007）。团头鲂低氧处理 4 或 7d 时，突出的薄片高度和鳃的平均薄片面积显著增加，这些变化导致 ILCM 高度和体积降低，然而恢复氧气 1 周后，鳃小片即可逆地嵌入 ILCM 细胞（Wu et al.，2017）。一般来说，ILCM 减少，导致层状表面积增加了暴露于水中的

第三章 洛氏鱥耐受短期低氧胁迫的生理机制

表 3-2 低氧处理后洛氏鱥鳃指数分析

项目	对照组	0.5h 3mg/L	0.5h 0.5mg/L	6h 3mg/L	24h 3mg/L	恢复组
突出薄片厚度(μm)	29.04±3.63[e]	12.32±2.26[c]	10.75±3.51[b]	9.24±2.82[a]	7.92±1.44[a]	16.04±2.92[d]
突出薄片高度(μm)	64.19±4.69[a]	65.76±6.14[a]	78.02±6.2[b]	83.05±6.85[c]	127.62±7.06[d]	66.14±9.77[a]
突出薄片的基础长度(μm)	162.09±7.19[a]	158.74±2.31[a]	159.24±0.63[a]	164.33±1.78[ab]	165.4±1.73[ab]	172.76±1.26[b]
层间细胞团高度(μm)	52.62±1.03[a]	27.28±0.66[b]	31.94±0.74[c]	31.18±0.77[c]	31.68±0.85[c]	32±0.93[c]
鳃小片间距离(μm)	19.93±0.60[a]	14.33±0.87[b]	20.93±5.27[c]	26.34±0.46[c]	32.25±0.72[d]	24.20±0.97[c]
层间细胞团体积(μm^3)	169 975.78±5 155.58[a]	62 079.36±903.05[b]	106 464.41±421.83[c]	134 977.76±1 463.13[c]	169 014.62±1 766.92[a]	133 802.38±975.21[c]
突出薄片面积(μm^2)	23 696.42±1 051.45[a]	23 289.84±338.79[a]	27 663.75±109.61[b]	30 363.72±329.14[c]	46 913.29±490.44[d]	25 567.35±186.35[e]

注:标有不同字母表示差异显著($P<0.05$),否则差异不显著。

面积，鱼类或许可以通过此方式提高氧气转移能力（Dolci et al.，2017；Wu et al.，2017）。金鱼鳃上层间细胞团的存在显著降低了功能性层状表面积并增加了气体转移的扩散距离，因此可能对呼吸气体的转移不利（Tzaneva et al.，2011）。层间细胞团的减少幅度与 LOE 呈负相关，与缺氧耐受性呈正相关（Dhillon et al.，2013）。Borowiec et al.（2015）研究表明减少的层状表面积可以减少离子损失，以及减少由增加的层状表面积和过高的渗透压引起的活性离子摄取。研究表明，低氧和运动会导致腹主动脉血压升高，从而导致柱细胞的空间改变（Fritsche and Nilsson，1993；Nilsson，1986）。因此可以得出结论，一个物种越活跃，它就越依赖功能表面积（FSA）的增加来满足增加的需氧量。不太活跃的物种很少通过依赖血流动力学调整来利用新陈代谢的上限（Gonzalez and McDonald，1994）。气体扩散距离减少是对缺氧的积极适应，本研究结果表明，洛氏鳄为面对低氧胁迫表现出活跃性的鱼类。

二、低氧处理后洛氏鳄肝组织形态学观察

（一）肝组织形态学观察

对照组肝脏呈现有规则的、排列紧密的肝细胞以及一些大液泡（彩图 4，表 3-3）。洛氏鳄的肝组织在 0.5h 3mg/L 低氧处理时，几个区域显示出明显的组织学变化，肝细胞面积与对照组相比显著增加。低氧处理组的肝细胞面积比对照组显著增加。然而，恢复氧气组的肝细胞面积与对照组相比没有显著差异。此外，0.5h 3mg/L、6h 3mg/L 和 24h 3mg/L 低氧组的肝细胞核面积均显著增加（$P<0.05$）。尽管如此，与对照组相比，0.5h 0.5mg/L 低氧处理组没有显著差异（$P>0.05$）。在 0.5h 0.5mg/L 低氧处理时，肝组织中出现大空泡。由于中度缺氧，空泡化逐渐明显，细胞轻微肿胀，特别是在 24h 3mg/L 时，而恢复氧气组肝中的空泡不明显。

表 3-3　肝细胞和细胞核面积（μm^2）

项目	对照组	0.5h 3mg/L	0.5h 0.5mg/L	6h 3mg/L	24h 3mg/L	恢复组
肝细胞面积	198.24±18.53[a]	306.41±15.64[d]	288.68±13.78[cd]	250.2±22.44[bc]	257.46±14.48[bc]	238.8±11.56[a]
肝细胞核面积	24.03±3.28[ab]	36.88±4.02[c]	15.95±2.77[a]	31.07±4.28[b]	34.98±3.48[b]	34.98±3.48[b]

注：标有不同数值表示差异显著（$P<0.05$），否则差异不显著。

（二）综合分析

肝脏参与包括葡萄糖和脂质稳态在内的多种代谢过程，对于维持鱼类应激平衡具有重要作用，也是受环境条件影响较为明显的器官之一（Jin et al.，

2018；Polakof et al.，2012）。有研究分析发现，长时间缺氧后肝脏会坏死，并显示出中度至重度脂质肝细胞空泡化区域（Mustafa et al.，2012）。大盖巨脂鲤在缺氧条件下表现出高比率的肝损伤、细胞空泡化和局灶性坏死，导致组织损伤以及基因表达和酶活性改变（Silva et al.，2019）。然而目前关于低氧胁迫后鱼肝的变化存在一些相互矛盾的结果，这可能与物种对缺氧耐受性的差异或低氧胁迫时间的差异有关。与尼罗罗非鱼的结果类似（Li et al.，2018），我们的结果表明，对照组的洛氏鱥含有较大的空泡，而缺氧鱼的空泡减小，说明低氧胁迫导致洛氏鱥脂质降解。

第三节　低氧胁迫对洛氏鱥肝脏糖脂代谢能力的影响

在动物中，糖代谢是能量获取的主要途径，尤其是在应激环境条件中，糖是胁迫后生物被动使用的第一能量物质。本研究分析了糖代谢酶的活性，包括乳酸脱氢酶（lactate dehydrogenase，LDH）、磷酸果糖激酶（phosphofructokinase，PFK）、丙酮酸激酶（pyruvate kinase，PK）、己糖激酶（hexokinase，HK）、琥珀酸脱氢酶（succinate dehydrogenase，SDH）和6-磷酸葡萄糖脱氢酶（6-phosphate-glucose dehydrogenase，G6PDH）（图3-2）。在本研究中，乳酸脱氢酶在低氧处理（0.5h 3mg/L）时上升，但是在极低氧气浓度胁迫（0.5h 0.5mg/L）时下降，在24h 3mg/L胁迫时也表现为下降，在恢复氧气浓度组乳酸脱氢酶活性最低（$P<0.05$）。磷酸果糖激酶在低氧处理期间酶活性增加，但是没有显著性变化（$P>0.05$）。丙酮酸激酶在6h 3mg/L处理组的酶活性显著增加（$P<0.05$），但是在其他组没有显著性变化（$P>0.05$）。己糖激酶在0.5h 3mg/L，6h 3mg/L和24h 3mg/L低氧处理时，酶活性显著降低（$P<0.05$）。琥珀酸脱氢酶在恢复氧气浓度组酶活性最高。6-磷酸葡萄糖脱氢酶活性在24h 3mg/L低氧胁迫时显著降低（$P<0.05$），其他组中变化不显著（$P>0.05$）。

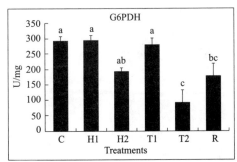
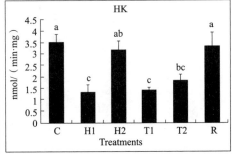

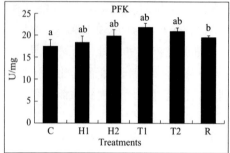
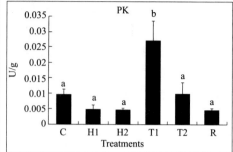
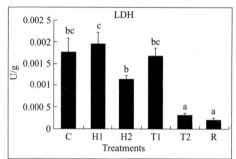
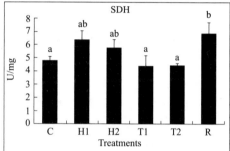
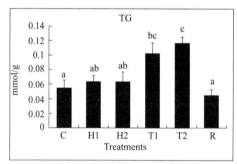
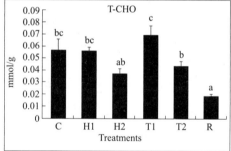
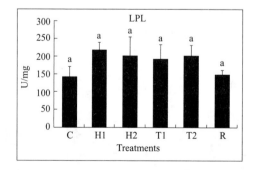

图 3-2 低氧胁迫下洛氏鱥糖脂代谢酶活性

注：其中 C 为对照组，H1 为 0.5h 3mg/L 组，H2 为 0.5h 0.5mg/L 组，T1 为 6h 3mg/L 组，T2 为 24h 3mg/L 组，R 为恢复组。其中不同字母代表差异显著（$P<0.05$）。

第三章　洛氏鱥耐受短期低氧胁迫的生理机制

（一）低氧处理对洛氏鱥脂代谢相关活性的影响

为了解低氧处理过程中洛氏鱥的脂质和氨基酸代谢的作用，本研究分析了洛氏鱥甘油三酯（triglyceride，TG）、总胆固醇（total cholesterol，T-CHO）、脂蛋白脂肪酶（lipoprotein lipase，LPL）在短期低氧胁迫后的变化。研究结果表明，甘油三酯在低氧胁迫24h 3mg/L时浓度最高。总胆固醇在6h 3mg/L时浓度最高。

（二）低氧处理对洛氏鱥氨基酸代谢相关活性的影响

本研究中，同时测定了短期低氧胁迫下洛氏鱥总氨基酸（total amino acid，T-AA）、谷氨酸脱氢酶（glutamate dehydrogenase，GDH）和谷草转氨酶（GOT）的变化。经过短期低氧胁迫后，洛氏鱥中谷氨酸脱氢酶和谷草转氨酶显著下降（$P<0.05$）。总氨基酸含量在0.5h 0.5mg/L，6h 3mg/L和24h 3mg/L组上升，0.5h 0.5mg/L和恢复低氧组下降（图3-3）。

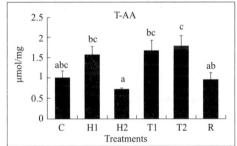
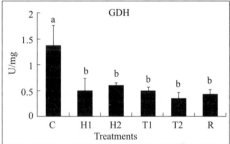
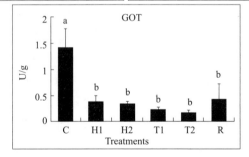

图3-3　低氧胁迫下洛氏鱥氨基酸代谢酶活性

注：其中C为对照组，H1为0.5h 3mg/L组，H2为0.5h 0.5mg/L组，T1为6h 3mg/L组，T2为24h 3mg/L组，R为恢复组。其中不同字母代表差异显著（$P<0.05$）。

（三）综合分析

环境低氧的压力总是会激活糖酵解以满足能量需求（Gracey et al.，2011；Richards，2011；Speers-Roesch et al.，2009）。通常厌氧糖酵解可能是急性低氧应激下的主要能源，急性低氧应激后糖异生被抑制（Richards，2011）。在缺氧条件下，葡萄糖分解为丙酮酸和少量 ATP，然后丙酮酸在乳酸脱氢酶的催化下被还原为乳酸（Connett et al.，1990；Mann，1970；Serebrovska et al.，2019）。本研究结果表明，洛氏鱥经低氧胁迫后，磷酸果糖激酶和丙酮酸激酶活性显著增加，这种变化仅发生在 6h 3mg/L 组，而乳酸脱氢酶活性在 0.5h 0.5mg/L 低氧处理组中略有增加。这与大黄鱼（*Larimichthys crocea*）低氧结果类似，即糖酵解的关键酶（如 HK，PK 和 PFK）以及与糖异生相关的酶（如 G-6-Pase）在短期低氧胁迫时显著上调（Ding et al.，2020）。在大口黑鲈（*Micropterus salmoides*）研究中发现，低氧会降低葡萄糖和糖原水平，增加乳酸脱氢酶活性，并显著降低肝脏中的总 ATP 含量（Rye and Lamarr，2015）。磷酸果糖激酶和丙酮酸激酶是糖酵解限速酶，本研究中，洛氏鱥增加的磷酸果糖激酶和丙酮酸激酶活性也揭示了洛氏鱥在短期中度缺氧的情况下厌氧代谢的增强（Johnston and Bernard，1982；Martinez et al.，2005）。除了琥珀酸脱氢酶之外，我们的结果表明其他酶在洛氏鱥的糖代谢途径中没有发挥重要作用，表明它们的催化活性受到抑制。琥珀酸脱氢酶是有氧代谢中唯一嵌入线粒体内膜的酶（Hägerhäll，1997），是有氧呼吸的限速酶（Ding et al.，2019）。本研究表明，琥珀酸脱氢酶在急性缺氧组（0.5h 0.5mg/L 和 0.5h 3mg/L）和复氧组中显著升高，说明有氧呼吸在此阶段发挥了作用。综上所述，糖酵解和有氧代谢均在洛氏鱥短期胁迫内（6h 内）起作用。

此外，大量研究表明，鱼类在急性缺氧期间的主要能量来源是无氧代谢，而不是脂质分解。例如，Ding et al.（2019）研究发现大黄鱼肝脏中甘油三酯和总胆固醇水平下降，脂蛋白脂肪酶的活性没有明显变化。再如尼罗罗非鱼急性缺氧应激时，肝糖原和肌肉糖原的含量明显降低，但甘油三酸酯（TG）没有显著差异（Li et al.，2018）。但是，大口黑鲈的研究发现低氧胁迫会增加肝脏中甘油三酯的动员和脂质过氧化作用：大口黑鲈中的甘油三酯浓度迅速升高，与缺氧的严重程度呈正比，并在缺氧应激下 24h 后达到峰值（Sun et al.，2020），这与本研究结果相似。脂蛋白脂肪酶催化甘油三酯水解为甘油和游离脂肪酸（Augustus et al.，2003；Westerterp et al.，2006），本研究结果显示 6h 3mg/L 和 24h 3mg/L 的甘油三酯和脂蛋白脂肪酶水平较高。因此，笔者推测短期低氧期间，能量也可以由脂质代谢提供。

碳水化合物代谢在能量供应中起着至关重要的作用，而氨基酸代谢是应付急性低氧应激的重要支持特征（Ding et al.，2020）。无论硬骨鱼是否迁移，当能量供应不足时，蛋白质会优先或最终转化为碳水化合物（Butier et al.，

1969）。红鲑（*Oncorhynchus nerka*）研究表明，当所有其他底物都耗尽时，蛋白质将在低氧后期成为主要的能量物质（Mommsen et al.，1980）。谷氨酸脱氢酶，涉及氨基酸代谢和碳水化合物代谢（Scott-Taggart et al.，2002）。已有研究表明严重缺氧后果可能是在低氧条件下抑制谷氨酸脱氢酶活性，从而关闭氨基酸代谢和碳水化合物代谢途径之间的这个关键连接点（Dawson and Storey，2012），促进氨基酸在厌氧途径下通过脱氨作用作为能量来源（Boyko et al.，2012；Gottlieb et al.，2003）。谷草转氨酶是必需氨基酸的合成和蛋白质代谢的三个关键酶之一（Sinha et al.，2013）。研究表明，鲫在缺氧条件下会抑制缺氧后充氧过程中的谷草转氨酶活性。低氧应激条件下，军曹鱼谷草转氨酶活性增加，然而复氧后24h谷草转氨酶活性显著低于对照组，这是由于复氧后短期内持续氧合对肝细胞造成了损伤（Huang et al.，2021）。洛氏鱥在缺氧处理期间谷氨酸脱氢酶和谷草转氨酶的酶活性降低，这表明氨基酸和碳水化合物代谢的联合作用对缺氧反应不明显。在大黄鱼的研究中，发现24h低氧胁迫后，谷草转氨酶和谷氨酸脱氢酶显著上调（Ding et al.，2020）。但是在本研究中，洛氏鱥经过短期低氧胁迫后，肝脏中谷氨酸脱氢酶和谷草转氨酶显著下降，这表明蛋白质代谢发挥的作用具有物种差异性。

第四节 低氧对洛氏鱥肝脏和鳃抗氧化能力和非特异性免疫能力的影响

水生生物在进化过程中形成了一套自我保护的抗氧化体系。在低氧胁迫时，体内的抗氧化体系发挥作用进行自我保护。缺氧胁迫下，鱼体可通过各种抗氧化酶，如超氧化物歧化酶（superoxide dismutase，SOD），过氧化氢酶（catalase，CAT）和丙二醛（malondialdehyde，MDA）等清除多余的氧自由基，导致鱼体的抗氧化机制发生相应变化（Pichavant et al.，2002）。酸性磷酸酶（acid phosphatase，ACP）和碱性磷酸酶（alkaline phosphatase，AKP）是水生生物中两种重要的磷酸酶，它们参与降解外来蛋白质、碳水化合物和脂质（Liu et al.，2004）。酸性磷酸酶通常用作检测细胞组分中溶酶体的标记物，也可用作评估环境污染的可靠工具（Rajalakshmi et al.，2005），碱性磷酸酶是存在于几乎所有动物细胞膜中的内在质膜酶（Jing et al.，2006）。本研究经短期低氧胁迫后，检测洛氏鱥肝和鳃的抗氧化能力和非特异性免疫相关酶活性变化。

（一）低氧处理对洛氏鱥肝脏抗氧化能力的影响

如图3-4所示，超氧化物歧化酶在0.5h 0.5mg/L，0.5h 3mg/L和6h

3mg/L 组都表现为显著增加（$P<0.05$），但是 CAT 酶活性在低氧胁迫时显著下降（$P<0.05$），此外 MDA 在所有低氧组和复氧组均表现为显著下降（$P<0.05$）。

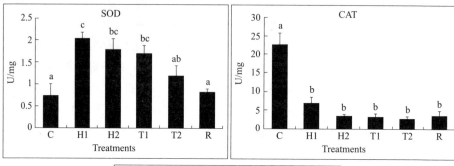

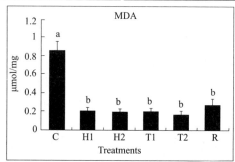

图 3-4 低氧胁迫下洛氏鱥肝脏抗氧化酶活性

注：其中 C 为对照组，H1 为 0.5h 3mg/L 组，H2 为 0.5h 0.5mg/L 组，T1 为 6h 3mg/L 组，T2 为 24h 3mg/L 组，R 为恢复组。其中不同字母代表差异显著（$P<0.05$）。

（二）低氧处理对洛氏鱥鳃抗氧化能力的影响

如图 3-5 所示，经低氧处理后，与对照组相比，鳃超氧化物歧化酶在 0.5h 0.5mg/L 组升高，恢复组活性增加（$P<0.05$）。过氧化氢酶除在 6h 3mg/L 组外，均比对照组升高（$P<0.05$）。此外，丙二醛 MDA 在 0.5h 0.5mg/L 组显著增加（$P<0.05$），但是在其他组中变化不显著（$P>0.05$）。

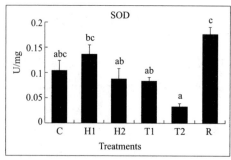

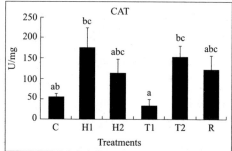

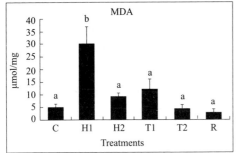
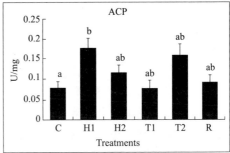
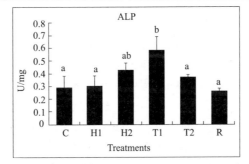

图 3-5 低氧胁迫下洛氏鱥鳃 SOD、CAT、MDA、ACP 和 ALP 酶活性

注：其中 C 为对照组，H1 为 0.5h 3mg/L 组，H2 为 0.5h 0.5mg/L 组，T1 为 6h 3mg/L 组，T2 为 24h 3mg/L 组，R 为恢复组。

（三）低氧处理对洛氏鱥鳃的非特异性免疫能力的影响

经过短期低氧胁迫后，与对照组相比，酸性磷酸酶在 0.5h 0.5mg/L 组显著增加（$P<0.05$），但是其他组中变化不显著（$P>0.05$）。与对照组相比，碱性磷酸酶在 6h 3mg/L 显著增加（$P<0.05$），其他组中变化不显著（$P>0.05$）。

（四）综合分析

缺氧会在需氧生物的细胞内导致活性氧（reactive oxygen species，ROS）水平升高，如果 ROS 超过内源性抗氧化剂的淬灭能力，则会发生氧化应激（Rocha-Santos et al.，2018）。生物通常通过显著提高超氧化物歧化酶（SOD）和过氧化氢酶（CAT）等抗氧化酶的活性水平，以保护暴露于缺氧或应激因素的身体。鱼类经低氧处理后，抗氧化酶活的变化与物种和低氧处理时间有很大关系。例如低氧处理后，军曹鱼肝脏中超氧化物歧化酶活性水平下降，这是由于超氧化物歧化酶和过氧化氢酶对溶解氧变化的反应速度不同所致，认为缺氧暴露后，军曹鱼自行适应缺氧应激（Huang et al.，2021）。而大口黑鲈的相关研究表明在缺氧暴露下过氧化氢酶显著增加，而超氧化物歧化酶略有变化（Sun et al.，2020）。大黄鱼的研究表明低氧处理后的肝脏超氧化物歧化酶活性显著增加，这与本研究获得的结果一致（Wang et al.，2017a）。本研究中，低氧处理时，在氧化应激反应中，洛氏鱥的肝脏中超氧化物歧化酶比过氧化氢酶起着更重要的作用。而洛氏鱥的鳃中，超氧化物歧化酶在恢复氧气组活性显著增加，这与在斑点叉尾鮰的幼鱼（Wang et al., 2017）和另一种耐缺氧鱼金曼龙（*Trichogaster microlepis*）中的研究一致（Huang et al., 2015），表明洛氏鱥在再充氧后对氧化应激有强烈反应。此外，这些酶在洛氏鱥中的激活表明它们在能量供应和细胞保护中起着至关重要的作用，并进一步表明脂质代谢在增强缺氧驯化中起着至关重要的作用。

丙二醛（MDA）是自由基氧化后产生的，丙二醛的含量是监测机体受到脂质和细胞过氧化损伤的指标（Johannsson et al., 2018）。然而，在相对长期缺氧暴露下，与对照组相比，洛氏鱥的丙二醛含量和抗氧化酶活性没有显著增加，表明洛氏鱥对缺氧环境具有适应策略。在军曹鱼的研究中，缺氧胁迫下血清中的丙二醛含量降低，氧化损伤程度降低，推测是由于超氧化物歧化酶和过氧化氢酶消除部分缺氧状态的氧自由基，从而减少脂质过氧化损伤（Huang et al., 2021）。低氧产生过量的 ROS 产生会增加脂质过氧化，从而导致 MDA 产生。MDA 的持续存在可以通过分解 DNA、蛋白质和细胞质来破坏生物细胞（Yao et al., 2010）。洛氏鱥经短期极端缺氧处理后，鳃中的丙二醛含量比对照组显著增加（$P<0.05$）。在黄颡鱼幼鱼的研究中发现，肝脏和鳃中丙二醛含量与急性缺氧应激后溶解氧含量呈显著负相关，与大脑相比，肝脏和鳃是最早受到损伤的组织（Wang et al., 2021）。

本研究结果表明，洛氏鱥鳃组织中，低氧处理 0.5h 3mg/L 胁迫后酸性磷酸酶活性比对照组显著升高（$P<0.05$），而 6h 3mg/L 胁迫后碱性磷酸酶活性比对照组显著升高（$P<0.05$），说明碱性磷酸酶是在应激后期发挥作用。低氧处理会刺激中华绒螯蟹（*Eriocheir sinensis*）鳃中 ACP 和 AKP 的酶活性（Bao et al., 2020），表明 ACP 和 AKP 通过水解作用在免疫反应中发挥防御作用。本研究结果表明，低氧处理后，洛氏鱥鳃中酸性磷酸酶和碱性磷酸酶活性以及超氧化物歧化酶和丙二醛含量与相对长期缺氧暴露高度相关。

第五节 低氧处理后洛氏鱥肝脏中脂质代谢相关基因表达的变化

（一）低氧处理后洛氏鱥脂肪生成和脂肪分解相关基因表达量的变化

如图 3-6 所示，*ACSL4* 基因在 6h 3mg/L 处理时表达量最高，显著高于对照组（$P<0.05$），而在 0.5h 3mg/L 时表达量最低，但是与对照组相比不显著（$P>0.05$）。*FASN* 基因在 6h 3mg/L 处理时表达量最高，显著高于对照组和其他各组（$P<0.05$）。*PPAR* 基因在 6h 3mg/L 处理时表达量最高，显著高

于对照组和其他各组（$P<0.05$）。$PGC-1\alpha$ 基因在 6h 3mg/L 处理时表达量最高，显著高于对照组和其他各组（$P<0.05$）。$PGC-1\beta$ 基因在 6h 3mg/L 处理时表达量最高，显著高于对照组和其他各组（$P<0.05$）。

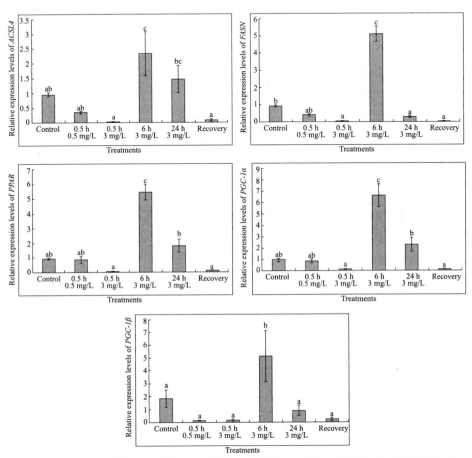

图 3-6 低氧暴露下洛氏鱥 $ACSL4$、$FASN$、$PPAR$、$PGC-1\alpha$ 和 $PGC-1\beta$ 的基因表达量
注：图中标有不同字母表示差异显著（$P<0.05$），否则差异不显著。

（二）综合分析

许多关键酶和转录因子在脂肪生成和脂肪分解中起着关键作用。为了评估调节脂质代谢的分子机制，以了解鱼类的缺氧反应，笔者分析了长链脂肪酸辅酶 A 连接酶 4（$ACSL4$）、过氧化物酶体增殖物激活受体-γ 辅激活因子 1（$PGC-1$）和脂肪酸合酶（$FASN$）在洛氏鱥低氧处理后的基因表达模式。过氧化物酶体增殖物激活受体伽马辅激活因子家族（PGC）有两个家族成员，即 $PGC-1\alpha$ 和 $PGC-1\beta$（Song et al.，2020）。$PGC-1\alpha$ 是一种对代谢和炎症具有深远而广泛影响的必需蛋白质，可调节各种生物作用，包括线粒体生物发生、脂肪酸氧化和葡萄糖代谢以及全身能量（Heshmati et al., 2020；Nierenberg et al., 2018）。有研究表明 PGC-1α 与棕色脂肪分化、产热和肌肉对啮齿动物运动的适应有关（Baar et al., 2002）。代谢稳态和底物利用的变化由各种转录因子调控，而 PGC-1α 是其中重要的调控因子，如作为代谢调节剂，多纹黄鼠（*Spermophilus tridecemlineatus*）骨骼肌中 *PGC-1α* 基因的表达在其冬眠时增加（Eddy et al., 2005）。在本研究中，洛氏鲅的 *PGC-1α* 基因表达水平在 6h 3mg/L 和 24h 3mg/L 低氧处理组增加，*PGC-1β* 表达水平也在 6h 3mg/L 低氧处理组增加。在亚洲高原地区特有鱼类齐口裂腹鱼的研究中，低氧 12h 后肝脏 *PGC-1α* 基因表达水平在 3mg/L 和 1.2mg/L 组均升高，揭示缺氧应激后脂质代谢增强以满足能量需求（Ma et al., 2015；Zhao et al., 2020）。PGC-1β 在肝脏脂质稳态和心脏功能中起重要作用，通常 PGC-1β 调节细胞分化和甘油三酯的积累（Lelliott et al., 2006；Song et al., 2020）。洛氏鲅在 6h 3mg/L 中 *PGC-1β* 表达量的增加与甘油三酯浓度的变化趋势一致，这证实 PGC-1β 参与了甘油三酯合成并与低氧暴露下的能量供应有关。此外，脂肪酸合酶是合成内源性脂肪酸的关键酶，也是甘油三酯从头合成中的关键酶（Ma et al., 2013）。长链脂肪酸-CoA 连接酶（ACSL）在脂肪酸（FA）代谢中起着至关重要的作用，其中 ACSL 通过将游离的长链 FA 转化为酰基-CoA 来催化初始步骤（Mashek et al., 2007）。在本研究中，脂肪酸合酶和 *ACSL4* 在 6h 3mg/L 低氧处理组的肝脏中表现出较高的表达水平。许多研究证明 *ACSL4* 在花生四烯酸和脂质代谢中发挥重要作用（Cao et al., 2000；Doll et al., 2016）。上述结果表明，缺氧应激可增强肝脏中脂肪酸的活化和转运。

第四章　洛氏鲅持续低氧和昼夜低氧处理后鳃和肝脏生理应答机制比较

　　水体缺氧被认为是淡水和海洋生态系统面临的最严重威胁之一，并且水体低氧与人为活动导致的营养过剩是密切相关的（Breitburg et al.，2018）。水体缺氧可能最终导致鱼类栖息地质量的下降，鱼类空间分布减少，鱼类数量和种类减少，甚至导致急性死亡事件（Pothoven et al.，2009；Roberts，2010；Vaquer-Sunyer and Duarte，2008）。在水生环境中，昼夜循环低氧是一种常见现象，产生的原因是氧分压（PO_2）由于白天的光合作用活动增加，而在夜间减少（D'Avanzo and Kremer 1994；Graham，1990；Williams et al.，2019）。昼夜循环低氧在春季和夏季尤为普遍（Stierhoff et al.，2003），而由于冬季水长时间冻结，水体会呈现持续性缺氧。高纬度地区的鱼类普遍生活在冬季持续性低氧和其他季节可能出现的昼夜循环缺氧条件中（Mathias and Barica，1980）。这与生活在低纬度地区的鱼类不同，后者仅在春季和夏季面临昼夜循环缺氧（Yang et al.，2013）。黑龙江省是中国纬度最高的地区，具有独特的自然光环境，形成季节性光周期变化。分布在黑龙江省的鱼类是研究昼夜节律低氧的良好生物目标。

　　大量研究表明，鱼类采用多种行为、生理和生化策略来应对不利因素，并且已证明对缺氧的耐受性因物种、生命阶段和栖息地而异（Dan et al.，2014；Farrell and Richards，2009；Pollock et al.，2007）。测定氧消耗率（oxygen consumption rates，MO_2）和失去平衡的临界点（LOE）是评估水生动物代谢活动和缺氧耐受性的两种方法（Barnes et al.，2011；Chapman et al.，2002；Yang et al.，2011）。大量研究分析了环境因素对不同水生物种耗氧量的影响。例如，在中华倒刺鲃（*Spinibarbus sinensis*）中发现，常氧处理和循环低氧暴露均导致该鱼氧消耗率减少，而稳定的缺氧暴露导致该鱼的氧消耗率增加（Dan et al.，2014）。

　　目前研究表明，几种鲤科鱼类在缺氧暴露期间鳃形态发生了巨大变化，包括鲫（Sollid et al.，2003）、青海湖裸鲤（*Gymnocypris przewalskii*）（Matey et al.，2008）和鳙（*Aristichthys nobilis*）（Dhillon et al.，2013b）等。鳃重塑增强了鱼类从水中的氧气吸收，并与鱼类的缺氧耐受性有关（Sollid et al.，2003）。肝脏是通过调节肝细胞结构来适应环境条件变化的重要器官（Ding et al.，2020；Petersen et al.，2017）。尽管肝脏在缺氧条件下调节代谢的能力已得到充分报道，但很少有研究描述缺氧条件下的组织学变化。当对缺氧暴露的生理补偿开始时，葡萄糖和脂质代谢反应最为明显（Li et al.，2018）。在大口黑鲈缺氧4h期间，肝糖原、葡萄糖和丙酮酸含量显著降低，缺氧8h会增加脂质动员并抑制氧化（Sun et al.，2020）。碳水化合物代谢在急性缺氧应激的能量供应中起着至关重要的作用，而当鱼类长期缺氧时脂肪分解升高，表明长期缺氧应激期间的能量需求主要来自脂质分解代谢（Li et al.，2018）。上述研究表明不同环境条件下鱼类的碳水化合物和脂质代谢的调节存在显著差异。最近的研究表明，持续缺氧和昼夜循环低氧会影响鱼类的行为，而循环低氧提高了鱼类的低氧耐受性（Borowiec et al.，2015；Brady et al.，2009；Williams et al.，2019；Yang et al.，2013）。

生物钟参与了生物的昼夜循环活动，是一种内源性和光驱动的转录振荡器，负责生理、行为和新陈代谢的24h节律，使各种生物能够在与外部环境条件相结合的情况下维持其生命活动，以天为周期，且周期性通常与季节有关（Bottalico and Weljie，2021；Imaizumi，2010；Pittendrigh，1993），因此光周期受到季节和纬度的影响。在蓝藻的研究中，生物钟已被证明在光照水平较低的高纬度海洋中比在低纬度海洋中更能提高能源效率（Hellweger et al.，2020）。生物钟基因包括 Clock、Bmal1（brin and muscle arnt-like protein 1）、Per1（period 1）、Per2（period 2）、Per3（period 3）、Cry1（cryptochrome 1）、Cry2（cryptochrome 2）和其他生物钟基因等。这些基因一起形成一个反馈回路，调节中枢系统和器官的节律功能（Li et al.，2020）。Per 与 Cry 异源二聚化，易位至细胞核，通过阻断 CLOCK/BMAL 复合物来抑制 Per1、Per2、Per3、Cry1 和 Cry2 基因的转录，因此细胞质中的 Per 和 Cry 蛋白会在当天晚些时候耗尽（Tosini et al.，2008）。随着 Per 与 Cry 蛋白的减少，负调节因子对 Clock 和 Bmal1 基因转录的抑制消失，并通过正向调节因子刺激 Per 和 Cry 基因的转录（Barnes et al.，2003）。Betancor 等人（2020）阐明了大西洋蓝鳍金枪鱼（Thunnus thynnus）脂质代谢的时钟基因调控，证明所有6个时钟基因（Bmal1、Clock、Per1、Per2、Per3、Cry1 和 Cry2）在大脑和肝脏中都表现出节律性。转录时钟通过借助各种酶感知细胞的还原状态与氧化状态，与新陈代谢相关联（Edgar et al.，2012；Rey and Reddy，2013）。对三刺鱼（Gasterosteus aculeatus）的研究表明，Per1 的 mRNA 水平和乳酸脱氢酶活性在黑暗期之前趋于最低，但这种模式受到低氧环境的干扰（Prokkola et al.，2015）。

第一节　持续低氧和昼夜低氧处理后 MO_2 值和血液学参数的变化

在持续低氧、昼夜循环低氧和正常氧气组（对照组）中，耗氧率缓慢下降

第四章 洛氏鲅持续低氧和昼夜低氧处理后鳃和肝脏生理应答机制比较

(图 4-1)。当氧含量高于 3mg/L 时,正常氧气组的耗氧率明显低于其他两组 ($P<0.05$)。在氧气浓度 4~7.5mg/L 时,三组之间的 MO_2 值没有显著差异 ($P>0.05$)。然而,当氧气浓度在 0.45~2.4mg/L 范围内时,与正常氧气组和持续低氧组进行比较,昼夜循环低氧组鱼的耗氧率表现出显著性降低 ($P<0.05$)。此外,当氧气浓度从 2.4mg/L 降至 0.45mg/L 时,正常氧气组的耗氧率显著高于持续低氧组和昼夜循环低氧组 ($P<0.05$)。

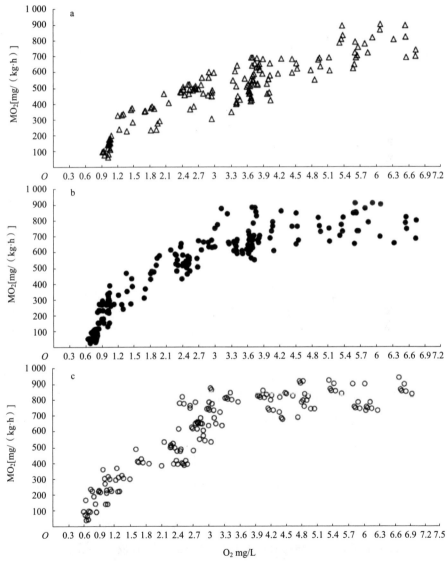

图 4-1 正常氧气组、持续低氧组和昼夜循环低氧组鱼的常规代谢率（MO_2）
a. 正常氧气组 b. 持续低氧组 c. 昼夜循环低氧组

（一）血液值变化

通常鱼的血液参数都会受到缺氧的影响，因此血液参数是评价鱼类外源性和内源性变化的重要生理指标。鱼的常规血液学评估包括测量红细胞（RBC）、血红蛋白（HGB）、血细胞比容（Hct）、平均红细胞血红蛋白浓度（MCHC）、平均红细胞体积（MCV）和平均红细胞血红蛋白含量（MCH）。如表4-1所示，与暴露于正常氧气环境的鱼相比，暴露于持续低氧组或昼夜循环低氧组的鱼的红细胞显著增加（$P<0.05$）。在缺氧处理下，持续低氧组鱼的血红蛋白在低氧第2天显著下降（$P<0.05$）。相反，除第2天和第6天外，昼夜循环低氧组的血红蛋白增加。昼夜循环低氧组第4天血细胞比容显著升高（$P<0.05$）。与对照组相比，缺氧时平均红细胞血红蛋白浓度显著升高，第8天变化最显著（$P<0.05$）。此外，与对照组相比，持续低氧组与昼夜循环低氧组的平均红细胞体积在第8天都表现出显著降低（$P<0.05$）。

（二）综合分析

本研究中，持续低氧和昼夜循环低氧处理下耗氧率逐渐减少，这说明洛氏鲅可以适应广泛的溶解氧范围，类似的结果也在中华倒刺鲃（*S. sinensis*）中观察到（Dan et al.，2014）。从生理上讲，一旦氧气浓度开始下降，鱼就会开始减少耗氧量。起初，这表现为标准代谢率降低（Lewis et al.，2007），而随着氧气浓度的降低，表现为代谢抑制等（Obirikorang et al.，2020）。当氧气水平范围为4~7.5mg/L时，耗氧率值在对照组、持续低氧组和昼夜循环低氧组比较接近，这也与在中华倒刺鲃中观察到的耗氧率是比较相似的（Dan et al.，2014），表明洛氏鲅正在适应这段时间内迅速减少的水体氧气浓度。在对尼罗罗非鱼（*O. niloticus*）的研究中，在昼夜循环低氧条件下鱼的耗氧率几乎没有受到影响，这表明鱼类可以依赖厌氧代谢机制，但是也只在一定的低氧范围内（Obirikorang et al.，2020）。洛氏鲅中，当氧气浓度低于2.4mg/L时，耗氧率下降。这种效应在昼夜循环低氧组中比在持续低氧组中更为明显，表明昼夜循环低氧可以增加洛氏鲅的缺氧耐受性。南方鲇（*Silurus meridionalis*）和虹鳟的研究均证实昼夜循环缺氧增强了其低氧耐受性（Williams et al.，2019；Yang et al.，2013）。有人认为，南方鲇在昼夜循环低氧后对缺氧的耐受性具有很大的可塑性，这可能与其生活环境中频繁发生的昼夜循环缺氧现象有关（Yang et al.，2013）。洛氏鲅生活在湍流浅水中，夏季环境循环缺氧，冬季也发生长时间持续缺氧。在本研究中，与持续低氧的鱼相比，昼夜循环低氧组的耗氧率下降得更明显。在虹鳟中的研究也证实昼夜循环低氧组触发能量代谢的保护性重塑，提高了缺氧耐受性（Williams et al.，2019）。本研究证实耐缺氧物种如洛氏鲅在应对这种环境压力时具有相当大的可塑性。

第四章 洛氏鲅持续低氧和昼夜低氧处理后鳃和肝脏生理应答机制比较

表 4-1 低氧处理后洛氏鲅血液参数测定

项目	对照组	持续低氧组					昼夜循环低氧组				
		持续低氧2天	持续低氧4天	持续低氧6天	持续低氧8天	持续低氧10天	昼夜循环低氧2天	昼夜循环低氧4天	昼夜循环低氧6天	昼夜循环低氧8天	昼夜循环低氧10天
RBC	1.64±0.09	1.88±0.21^A	2.51±0.24^{AB}	2.54±0.19^{B*}	2.56±0.17^{B*}	2.71±0.12^{B*}	1.78±0.25^a	3.07±0.71^b	2.69±0.15^{ab}	2.87±0.1^{ab*}	2.89±0.13^{ab*}
HGB	108.71±2.27	80.25±6.73*	101.75±11.32	95.25±12.48	104.5±11.23	111.5±9.13	86.67±4.91^{a*}	151±14.47^b	105.67±11.61^a	116.67±17.03^{ab}	114.67±8.11^{ab}
Hct	31.05±1.78	30±2.7	38.5±3.91	36.95±1.69	33.65±5.49	39.03±2.63	36.1±3.96^a	49.8±2.11^a	36.93±1.59^a	36.45±1.38^a	31±5.84^a
MCHC	257.59±3.04	267.75±9.31^A	263.75±11.48^A	259±8.22^A	317.5±17.48^{B*}	285.25±9.83^{AB*}	273±13.96	274.67±11.02	267.67±3.84	309.75±29.78*	275.5±6.12
MCH	42.37±0.55	41.17±2.49	42.6±2.02	41.5±0.84	43.7±2.27	46.57±2.94	49.47±2.37	44±0.8	43.75±2.63	50.43±6.91	43.1±1.59
MCV	163.78±1.32	160.88±6.26^B	154.25±7.77^{AB}	162.83±5.58^B	139.35±0.78^{A*}	155.95±6.68^{AB}	177.35±5.1^b	171.87±8.61^{ab}	158.43±8.91^{ab}	150.33±7.99^a	161.43±6.61^{ab}

注:大写字母为持续低氧处理组;小写字母为昼夜循环低氧处理组。* 表示与对照组有显著性差异($P<0.05$)。

在生理压力下，血液参数是鱼类外源性和内源性变化最重要的指标之一。血液学参数被用来评估血液的氧气携带能力，并受到环境条件的影响。因此，血液的携氧能力可能严重影响鱼在低氧含量栖息地中的生活。应激会影响鱼的许多血液参数，如血细胞比容、红细胞和血红蛋白水平，这些对生物的氧气吸收至关重要（Bao et al.，2018）。在本研究中，红细胞计数随着缺氧而增加，类似于先前在比目鱼（*Pleuronectes flesus*）（Soldatov，1996）、斑马鱼（*Danio rerio*）（Cypher et al.，2015）和真鲷（*Pagrus major*）（Nam et al.，2020）中的报告，表明缺氧会引发血液携氧能力的增加。红细胞的增加表明缺氧时合成能力增强，为生命机能提供额外的能量，这可能有助于获取更多的氧气并增加鱼中血液的气体运输能力（Wu et al.，2016）。同时，持续低氧和昼夜循环低氧暴露下，血红蛋白在第 2 天下降，并且在接下来的第 6 天、第 8 天和第 10 天之间没有显著差异，表明血红蛋白浓度在低氧初期没有明显的调节作用。有学者认为，红细胞的增加可能是由于脾脏收缩造成的（Affonso et al.，2002；Moura et al.，1997）。洛氏鱥中的血红蛋白浓度在第 4 天的循环缺氧中显著增加，证实对缺氧的生理和代谢适应是基于低氧处理持续时间的。在大盖巨脂鲤的研究中发现，缺氧条件下的血细胞比容和血红蛋白浓度在 96h 后下降，这可能是由于缺乏食物引起的禁闭或压力所致（Affonso et al.，2002）。在暴露于持续和循环缺氧 4 天后，洛氏鱥表现出血细胞比容的大幅增加。在虹鳟的研究中发现，伴随昼夜低氧处理，虹鳟逐渐采用更耐缺氧的表型，通过增加血细胞比容和平均红细胞血红蛋白浓度、减少耗氧率、重新激活合成代谢和大量上调伴侣蛋白水平来增加氧气的吸收和运输（Williams et al.，2019）。有研究表明，生活在缺氧沼泽中的鱼类比生活在正常氧含量湖泊中的鱼类具有更高的血红蛋白和血细胞比容，从而显示出更强的携氧能力（Chapman et al.，2002）。

第二节 持续低氧和昼夜低氧处理后洛氏鱥组织形态学观察

一、持续低氧和昼夜低氧处理后洛氏鱥鳃组织形态学观察

在我们的实验中，对照组的样品表现出正常的鳃形态（彩图 5 和图 4-2）。与对照组相比，除了第 2 天和第 4 天，暴露于持续低氧组的鱼的鳃小片厚度增加（图 4-3）。在第 4 天和第 6 天，昼夜循环低氧组中鱼的鳃小片厚度增加。与常氧状态下的鱼相比，除第 2 天的昼夜低氧组，其余低氧组的鱼伸出鳃小片的长度增加。与对照组相比，持续低氧组除第 6 天外伸出鳃小片基部的长度显著增加。昼夜循环低氧组第 8 天和第 10 天伸出鳃小片基部的长度也显著增加

第四章　洛氏鲅持续低氧和昼夜低氧处理后鳃和肝脏生理应答机制比较

($P<0.05$)。对于第 10 天暴露于昼夜循环低氧组的鱼，相邻鳃小片距离显著增加（$P<0.05$）。持续低氧组第 8 天和第 10 天相邻鳃小片距离均显著增加。与对照组相比，持续低氧处理第 4 天层间细胞团高度显著增加；昼夜低氧处理后层间细胞团高度在第 2 天、第 6 天和第 8 天显著增加。与对照处理相比，所有持续低氧组和昼夜循环低氧组的鱼在第 4 天、第 6 天和第 8 天的鱼鳃的薄片表面积显著增加（$P<0.05$）（图 4-4）。与昼夜循环低氧组相比，持续低氧组的鱼鳃的薄片表面积更大。层间细胞团体积在第 2 天、第 8 天和第 10 天在持续低氧组和昼夜循环低氧组的鱼中显著增加。在昼夜循环低氧组的鱼中，层间细胞团体积在第 4 天显著减少（$P<0.05$）。除第 8 天，昼夜循环低氧组的鱼的层间细胞团体积均高于持续低氧组。

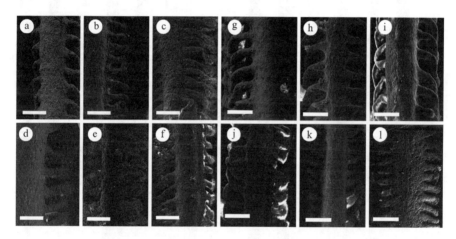

图 4-2　持续低氧和昼夜低氧处理后第 2 天、第 4 天、第 6 天、第 8 天和第 10 天洛氏鲅鳃的扫描电子显微照片（比例尺＝100μm）
a、g. 正常氧气组　b. 持续缺氧 2 天　c. 持续缺氧 4 天　d. 持续缺氧 6 天　e. 持续缺氧 8 天
f. 持续缺氧 10 天　h. 循环缺氧 2 天　i. 循环缺氧 4 天　j. 循环缺氧 6 天
k. 循环缺氧 8 天　l. 循环缺氧 10 天

二、持续低氧和昼夜低氧处理后洛氏鲅肝组织形态学观察

与正常氧气处理组（彩图 6）相比，第 2 天持续缺氧的鱼的肝细胞排列松散无序，部分细胞呈细胞质空泡化。持续低氧处理第 6 天，这些细胞排列松散无序，部分肝细胞核呈深染。此外，持续低氧处理后的第 10 天，肝细胞出现空泡的程度显著减小。相反，昼夜循环低氧组的肝脏样本在后期表现出明显的空泡化。此外，与对照组相比，持续低氧组肝脏切片的肝细胞和细胞核面积减少，在第 2 天达到最低水平（图 4-5）。在昼夜循环低氧中，这现象发生得较

晚。此外，10天昼夜循环低氧暴露后与对照组相比，肝细胞和肝细胞核面积显著降低（$P<0.05$）。

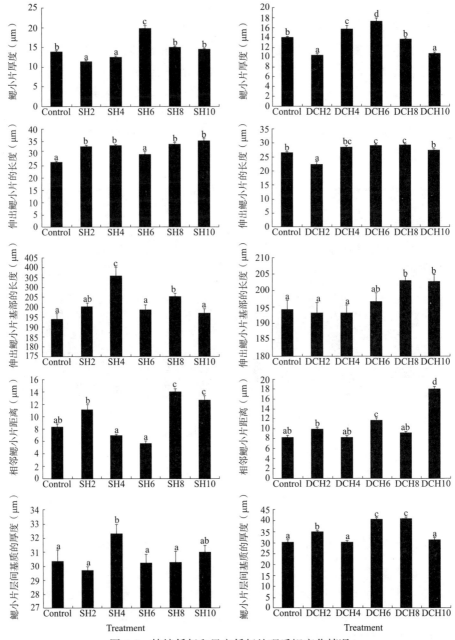

图 4-3 持续低氧和昼夜低氧处理后鳃变化情况

注：Control 表示对照组，SH2～10 表示持续低氧 2～10 天；DCH2～10 表示昼夜低氧处理 2～10 天。图中标有不同字母表示差异显著（$P<0.05$），否则差异不显著。

第四章　洛氏鳅持续低氧和昼夜低氧处理后鳃和肝脏生理应答机制比较

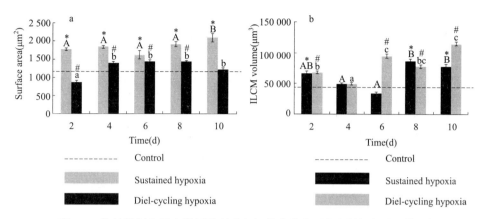

图 4-4　持续低氧和昼夜低氧处理后鱼鳃的薄片表面积和层间细胞团体积

a. 鳃表面积　b. 层间细胞团体积

注：＊表示与持续低氧组内的对照有显著差异，♯表示与昼夜低氧组内的对照有显著差异。大写字母为持续低氧处理，小写字母为昼夜低氧处理。图中标有不同字母表示差异显著（$P<0.05$），否则差异不显著。

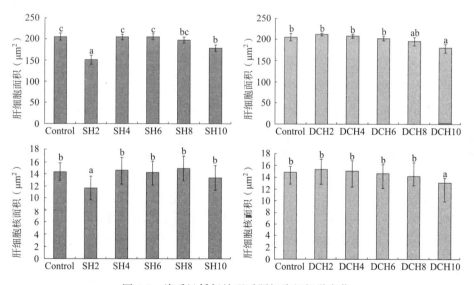

图 4-5　洛氏鳅低氧处理后肝细胞组织学变化

注：标有不同字母表示差异显著（$P<0.05$），否则差异不显著。

三、综合分析

在10天的处理过程中，笔者发现洛氏鳅伸出鳃小片基部的长度略有增加，同时低氧10天后鳃的表面积略有增加。此外，持续低氧的鱼的鳃薄片表面积比昼夜循环低氧条件下的鱼增加得更多，这可能是因为鳃受到持续低氧的持续影响。作为氧气传输级联的第一步,鳃组织对于鱼类氧气的吸收至关重要,因此,鳃表面积增加表明缺氧适应时,鱼类增加了氧气吸收(Borowiec et al.,2015;Evans et al.,2005;Hughes,1966)。然而,鳃的大气体交换面积促进了低渗环境中的被动离子损失,这需要主动离子泵来维持离子稳态(Borowiec et al.,2015)。一种看法认为,介导鳃中被动离子损失的机制是层间细胞团细胞的减少(Dhillon et al.,2013a;Nilsson et al.,2012)。本研究表明,在持续低氧 4 天时,层间细胞团体积显著减少。有研究表明,在低氧条件下鱼鳃表面积的减少会最大限度地减少离子破坏而不是促进氧气吸收(De Boeck et al.,2013;Matey et al.,2011;McDonald and McMahon,1977)。本研究结果证实,鱼对持续缺氧的反应比对循环缺氧的反应更严重。有趣的是,当鱼长时间暴露于缺氧环境中时,鳃小片之间的距离并没有显著增加(Yang et al.,2021)。有报道表明,鱼类对急性和长期缺氧表现出不同的反应机制(Douxfils et al.,2012;Jimenez et al.,2019;Sun et al.,2020;Wu et al.,2020),而这种差异响应与暴露时间有关。笔者发现,持续低氧组的鳃小片面积比昼夜循环低氧组中增加得更多,但昼夜循环低氧组的鱼,其层间细胞团(ILCM)体积比在持续低氧组中增加得更多。

在硬骨鱼肝脏中,肝脏中组织学变化可以表现在细胞质空泡化、细胞肥大、细胞变形、细胞核空泡化和细胞核变性等(Silva et al.,2019;Wang et al.,2017)。与本研究结果类似,在鱼类杀虫剂和低氧处理中都注意到细胞变形(Silva et al.,2019)。在低氧处理的大盖巨脂鲤肝脏中观察到细胞核空泡化和细胞核变性(Silva et al.,2019),但在洛氏鳄中未发现,这在一定程度上证明洛氏鳄具有耐受低氧的能力。在本研究中,暴露于持续低氧和昼夜循环低氧后,肝脏空泡明显减少。本研究中,洛氏鳄的肝细胞面积在持续低氧第 4 天增加,表明鱼在低氧 4 天时表现出对环境变化的适应。然而,在第 8 天与第 10 天,持续低氧组肝细胞面积减少并且出现了肝空泡化现象,表明脂质的降解,证实了脂质在持续低氧条件下可作为重要的能量来源。与持续低氧后肝细胞面积的增加不同,昼夜低氧组肝细胞面积减少,呈现明显的空泡化,表明昼夜低氧条件下脂质降解更为明显。缩小的肝空泡可能是与进一步在体内积累能量有关。在本研究中,洛氏鳄在持续低氧胁迫下第 2 天和在昼夜循环低氧条件下第 10 天,肝组织中肝细胞面积和细胞核面积的减少,这可能是由于在缺氧初期的持续低氧组和后期缺氧的昼夜循环低氧组的肝组织中出现了明显的反应。

第三节　持续低氧和昼夜低氧处理后肝脏糖脂代谢酶及抗氧化酶的活性变化

一、持续低氧和昼夜低氧处理对洛氏鱥肝脏糖脂代谢能力的影响

（一）持续低氧和昼夜低氧处理对洛氏鱥糖代谢相关酶活性的影响

本工作开展了持续低氧和昼夜低氧处理对洛氏鱥糖代谢相关酶活性的影响，主要包括己糖激酶（hexokinase，HK）、磷酸果糖激酶（phosphofructokinase，PFK）、丙酮酸激酶（pyruvate kinase，PK）、乳酸脱氢酶（lactate dehydrogenase，LDH）和琥珀酸脱氢酶（succinate dehydrogenase，SDH）。如图4-6所示，经过持续低氧处理，PFK酶活性在各组间，以及各组与对照组相比，均没有显著性变化（$P>0.05$）。经昼夜低氧处理，HK酶活性在第8天显著升高，高于对照组和其他各组（$P<0.05$）（图4-7）。LDH酶活性在昼夜低氧处理过程中，各组与对照组相比，均没有显著性变化（$P>0.05$），但是在持续低氧处理第2天值显著高于处理第6天和第10天（$P<0.05$）。SDH的酶活性在持续低氧处理的大部分组内均低于对照组，在处理第6天时值最高，在处理第8天时值最低。SDH的酶活性在昼夜低氧处理第8天时值最大并且显著高于第2天、第6天和第10天（$P<0.05$）。

（二）持续低氧和昼夜低氧处理对洛氏鱥脂代谢相关指标的影响

为了获得洛氏鱥持续低氧和昼夜低氧处理脂代谢相关指标的变化情况，本研究测定了胆固醇（cholesterol，CHO）和甘油三酯（triglycerides，TG）的含量。如图4-6和图4-7所示，胆固醇含量在持续低氧和昼夜低氧处理后，各组与对照组相比均无显著差异，且各组间没有显著差异（$P>0.05$）。TG在持续低氧的各实验组，与对照组相比均无显著性变化，且各组间没有显著性变化（$P>0.05$）。在昼夜低氧处理后，第10天TG含量达到最大值并且显著高于第4天（$P<0.05$），其他各组之间无显著性变化（$P>0.05$）。

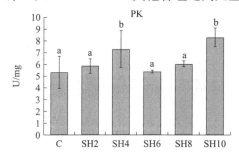
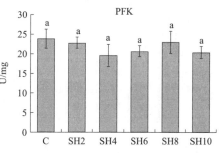

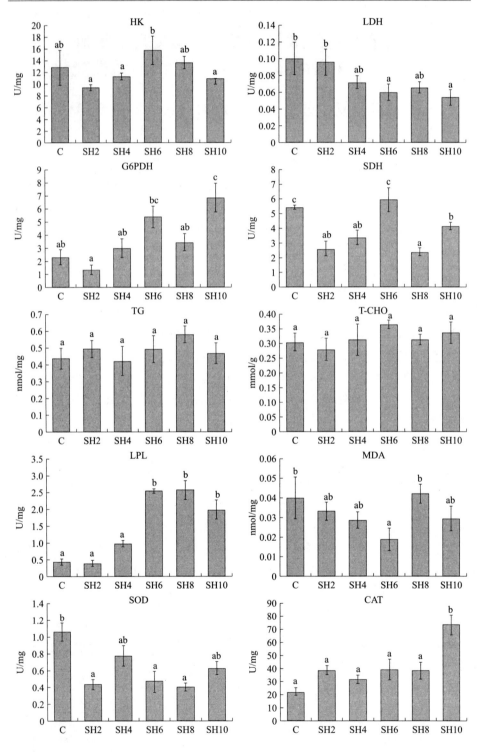

第四章 洛氏鱥持续低氧和昼夜低氧处理后鳃和肝脏生理应答机制比较

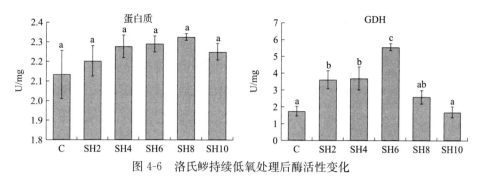

图 4-6 洛氏鱥持续低氧处理后酶活性变化

注：图中标有不同字母表示差异显著（$P<0.05$），否则差异不显著。C 表示对照组，SH2 表示持续低氧第 2 天，SH4 表示持续低氧第 4 天，SH6 表示持续低氧第 6 天，SH8 表示持续低氧第 8 天，SH10 表示持续低氧第 10 天。

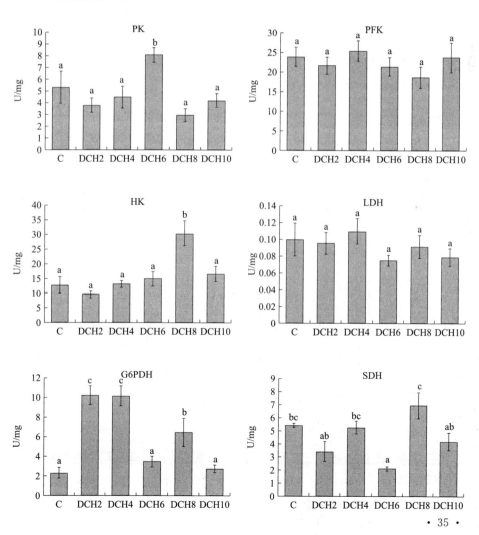

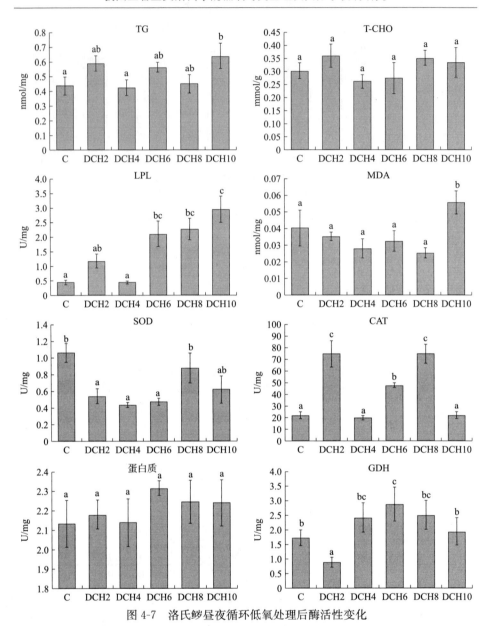

图 4-7 洛氏鱥昼夜循环低氧处理后酶活性变化

注：图中标有不同字母表示差异显著（$P<0.05$），否则差异不显著。C 表示对照组，DCH2 表示昼夜循环低氧第 2 天，DCH4 表示昼夜循环低氧第 4 天，DCH6 表示昼夜循环低氧第 6 天，DCH8 表示昼夜循环低氧第 8 天，DCH10 表示昼夜循环低氧第 10 天。

（三）持续低氧和昼夜低氧处理对洛氏鱥氨基酸代谢相关酶活性的影响

本研究测定了洛氏鱥持续低氧和昼夜低氧处理下谷氨酸脱氢酶（glutamate dehydrogenase，GDH）的活性。结果表明，GDH 在持续低氧处理时，除了第 8

天和第 10 天之外，各组均显著高于对照组（$P<0.05$），GDH 酶活性在第 6 天达到最大，显著高于其他各组（$P<0.05$）。GDH 在昼夜低氧处理时，第 6 天显著高于对照组和第 2 天、第 10 天处理组（$P<0.05$），且在第 6 天达到最大值。

（四）综合分析

在水生动物中，代谢调整允许生物体在环境低氧期间调动可用的能量储备（Gracey et al.，2011）。本项目首次开展了持续低氧和循环低氧对糖原代谢、脂类代谢和氨基酸代谢中酶活性的不同影响的研究，为研究鱼类在面对这些情况时产生能量的方式提供了信息。糖原代谢始终是能量获取的主要途径，尤其是在不稳定的环境中（Oliveira et al.，2004）。我们发现在持续低氧处理和昼夜低氧处理条件下，任何组的 PFK 都没有显著增加，而所有持续低氧组的 LDH 都减少，但在昼夜低氧处理期间没有显著变化。与对照组相比，第 6 天持续低氧处理鱼和第 8 天昼夜低氧处理鱼的 SDH 略有增加，并在所有其他缺氧组中减少。G6PDH 在低氧开始时显著增加，但在昼夜低氧处理期间的第 6 天和第 10 天没有明显变化，而在持续低氧处理的鱼中在第 6 天增加。因此，我们提出了这样的假设，即调节葡萄糖代谢不是洛氏鱥应对低氧暴露的唯一生理策略。有报告证实，糖酵解和糖异生的关键酶在短期低氧暴露期间显著上调（Ding et al.，2011）。相比之下，HK 和 PK 的低活性表明尼罗罗非鱼在慢性缺氧条件下糖酵解总体减少（Li et al.，2018）。有趣的是，我们还注意到第 6 天后昼夜低氧处理组的 HK 增加，并且在大多数持续低氧处理条件下 PK 活性增加。昼夜低氧处理的鱼中 HK 的轻微增加表明糖酵解的增强。PK 酶活性的升高意味着丙酮酸的产生，实现了糖、脂肪和氨基酸的相互转化，表明鱼类可能采用多种能量产生模式在持续低氧暴露期间维持生存。在虹鳟中，昼夜低氧处理减少了氧气消耗率，同时证明氧气转运级联效应有所改善，减少了低氧时对无氧代谢和糖原储存的依赖（Williams et al.，2019）。

洛氏鱥中的胆固醇浓度在两种低氧条件下没有显著差异。昼夜低氧处理的鱼中的甘油三酯（TG）浓度增加，尤其是在第 10 天，但在持续低氧处理的鱼中没有增加，这证实了昼夜低氧处理的鱼使用了甘油三酯，并且两种低氧模式下洛氏鱥在耐受策略方面存在差异。众所周知，甘油三酯是一种必不可少的能量来源，许多研究表明，在低氧压力下，甘油三酯浓度会显著增加，以帮助生物体维持能量平衡（Weber，2011）。然而，在虹鳟中，甘油三酯的浓度在昼夜循环低氧暴露下没有变化，说明虹鳟不使用甘油三酯作为能源（Williams et al.，2019）。虽然这两种鱼类都是高纬度冷水鱼类，但洛氏鱥总是在浅水区（湖泊、溪流等）中发现，因此面临着严重的持续和昼夜循环低氧条件。虹鳟大多生活在山间溪流的深水区（Magoulick and Wilzbach，1997），在那里利用垂直运动来避免周期性低氧环境。氨基酸代谢相关酶谷氨酸脱氢酶（GDH）

在持续低氧处理组中显著增加。这表明氨基酸代谢也参与了对持续低氧处理条件的响应,这与我们之前的理解一致,即鱼类在持续低氧的情况下使用更多形式的能量。

二、持续低氧和昼夜低氧处理对洛氏鲅肝脏抗氧化酶活性和脂质过氧化能力的影响

(一)持续低氧和昼夜低氧处理对洛氏鲅肝脏抗氧化能力的影响

首先分析了洛氏鲅响应持续低氧和循环低氧(昼夜循环低氧组)的总蛋白,笔者发现两种处理后都没有明显变化。与对照组相比,持续低氧组和昼夜循环低氧组的超氧化物歧化酶(SOD)均显著降低($P<0.05$)。在持续低氧条件下,过氧化氢酶(CAT)增加并在第 10 天时达到峰值。然而在昼夜循环低氧组中,除第 4 天和第 10 天外,CAT 水平显著增加($P<0.05$)。此外,在持续低氧组中,丙二醛(MDA)在第 8 天显著高于第 6 天处理组,并达到峰值;在昼夜循环低氧组中,它在第 10 天显著增加。

(二)综合分析

在本研究中,过氧化氢酶在第 10 天的持续低氧组和第 2 天、第 6 天和第 8 天的昼夜循环低氧组中显著增加。所有应激组的 SOD 均下降。MDA 在 10 天循环低氧时显著增加,并且在其他处理组之间没有显著差异。我们的结果表明 CAT 在持续低氧组和昼夜循环低氧处理时都发挥作用,可能是保护细胞在低氧期间免受损伤。一些鱼类在再氧化前通过增加不同抗氧化酶的活性、避免氧化应激或将其维持在生理上可控制的水平来调节它们的抗氧化系统(Hermes-Lima and Zenteno-Savín,2002)。过氧化氢酶活性在肝脏中发生了类似的增加,并且在金鱼缺氧暴露 8h 期间观察到,这表明该酶可能在减少氧化应激方面发挥作用(Lushchak et al.,2001)。

第四节 持续低氧和昼夜低氧处理后肝脏糖代谢相关基因和钟基因表达的变化

(一)持续低氧和昼夜低氧处理对洛氏鲅糖代谢相关基因表达的影响

本研究分析了 $Glut1$ 和 CCK 两个基因,结果如图 4-8 和图 4-9 所示,持续低氧第 8 天,$Glut1$ 基因表达结果显著上调($P<0.05$),明显高于对照组和其他组($P<0.05$),其他各组与对照组相比无显著差异($P>0.05$)。CCK 基因在持续低氧第 8 天显著上调($P<0.05$),达到最大值,其他各组与对照组相比无显著差异($P>0.05$)。在昼夜低氧处理第 6 天,$Glut1$ 基因表达结果显著上调($P<0.05$),明显高于对照组和其他组($P<0.05$),其他各组与对照组

相比无显著差异（$P>0.05$）。CCK 基因在昼夜低氧第 4 天和第 6 天显著上调（$P<0.05$），在第 4 天达到最大值，其他各组与对照组相比无显著差异（$P>0.05$）。

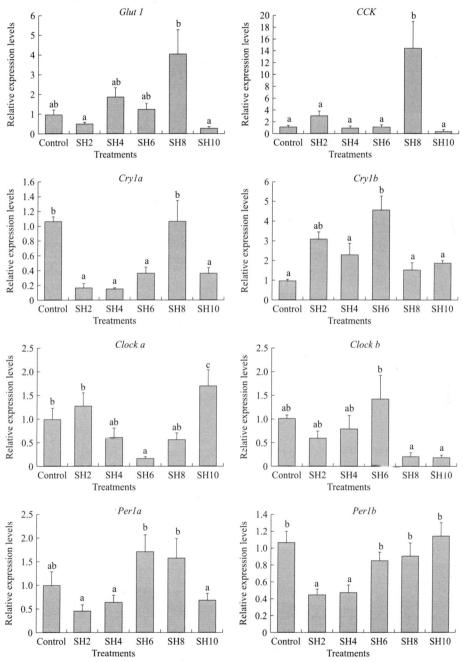

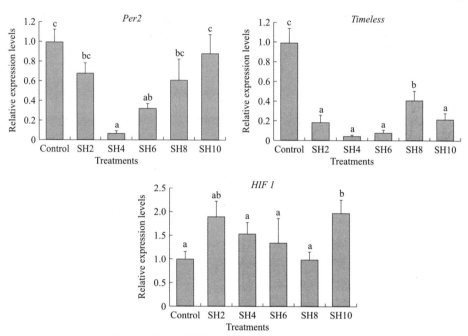

图 4-8 洛氏鱥持续低氧处理后基因表达量的变化

注：图中标有不同字母表示差异显著（$P<0.05$），否则差异不显著。

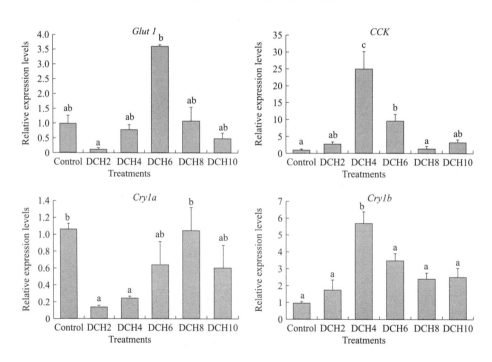

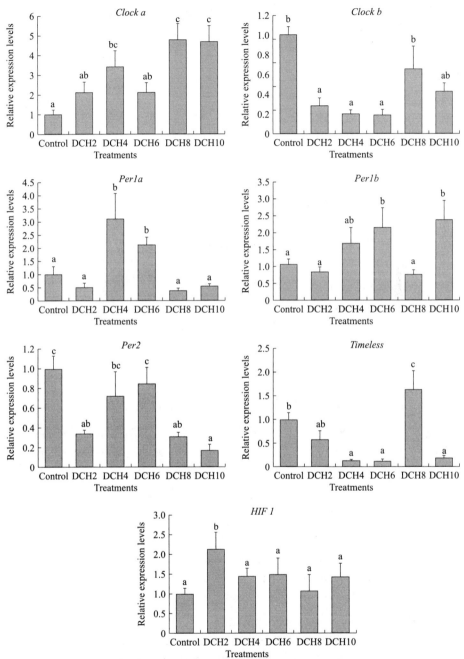

图 4-9 洛氏鱥昼夜低氧处理后基因表达量的变化

注：图中标有不同字母表示差异显著（$P<0.05$），否则差异不显著。

（二）持续低氧和昼夜低氧处理对洛氏鱥钟基因和 HIF 基因表达的影响

本研究分析了 $Cry1a$，$Cry1b$，$Clock\ a$，$Clock\ b$，$Per1a$，$Per1b$，$Per2$，$Timeless$ 和 $HIF1$ 等基因在持续低氧和昼夜低氧条件下的表达模式。结果表明，持续低氧第 8 天，$Cry1a$ 基因表达结果显著上调（$P<0.05$），明显高于其他组（$P<0.05$），但是与对照组相比无显著差异（$P>0.05$）。在昼夜低氧处理后，$Cry1a$ 基因表达呈现上升的趋势，在第 8 天表达量最高，显著高于第 2 天和第 4 天（$P<0.05$），其他缺氧处理组之间无显著差异（$P>0.05$）。$Cry1b$ 基因在持续低氧处理时仅在第 6 天显著上升，其余各组与对照组相比无显著差异（$P>0.05$）。$Cry1b$ 基因在昼夜低氧处理后，呈现先上升后下降的趋势，第 4 天表达量明显高于其他各组（$P<0.05$），且达到最高值，但是其他各组之间无显著差异（$P>0.05$）。$Clock\ a$ 基因在持续低氧处理时呈先下降后上升趋势，持续低氧第 10 天表达量最高，第 6 天表达量最低，其余各组与对照组相比无显著差异（$P>0.05$）。$Clock\ a$ 基因在昼夜低氧处理后，表达量在第 8 天和第 10 天显著高于第 2 天和第 6 天（$P<0.05$），但是与第 4 天没有显著差异（$P>0.05$），在第 8 天达到最高值，且第 4 天、第 8 天和第 10 天明显高于对照组（$P<0.05$）。$Clock\ b$ 基因在持续低氧第 6 天达到最大值，但是与对照组相比无显著差异（$P>0.05$）。$Clock\ b$ 基因在昼夜低氧处理后，仅第 8 天和第 10 天与对照组相比无差异，其余各组均显著低于对照组（$P<0.05$），在第 6 天表达量达到最低。$Per1a$ 基因在持续低氧第 6 天和第 8 天显著上调（$P<0.05$），高于对照组和其他各组，同时在第 6 天达到最大值。$Per1a$ 基因在昼夜低氧处理后，表达量在第 4 天和第 6 天显著高于对照组和其他各组（$P<0.05$），且在第 4 天为最高值。$Per1b$ 基因在持续低氧后表现为上升趋势，第 6 天、第 8 天和第 10 天显著高于其他低氧处理组（$P<0.05$），且在第 10 天达到最大值，但是表达量没有高于对照组。$Per1b$ 基因在昼夜低氧后，在第 6 天和第 10 天显著高于其他各组（$P<0.05$），在第 6 天和第 10 天显著高于对照组（$P<0.05$），在第 10 天达到最大值。在持续低氧处理下，除了第 10 天之外，$Per2$ 基因的表达量在第 4 天表达量最低，第 10 天最高。在昼夜低氧处理时，$Per2$ 基因各组均低于对照组，在第 10 天表达量达到最低。$Timeless$ 基因表达量在持续低氧处理 10 天内，第 8 天基因表达量最高，显著高于其他低氧处理组（$P<0.05$）。经过昼夜低氧处理后，$Timeless$ 基因表达量除第 2 天和第 8 天之外，均显著低于对照组（$P<0.05$），第 8 天表达量最高，显著高于其他各组（$P<0.05$）。$HIF1$ 基因的表达量在持续低氧条件下为先降低后升高模式，在低氧处理第 10 天表达量最高，在持续低氧处理第 10 天表达量显著高于对照组（$P<0.05$）。$HIF1$ 基因的表达量在第 2 天显著高于对照组（$P<0.05$），且第 2 天为最高值。

（三）综合分析

Glut1（葡萄糖转运蛋白1型）已知可调节细胞糖酵解和增殖（Xiao et al.，2018）。在这项研究中，与对照组相比，*Glut1* 的基因表达显著上调。在暴露于缺氧的大口黑鲈中也观察到了类似的反应，这表明鱼会增加葡萄糖的摄取以促进对这些压力源的反应（Yang et al.，2017）。胆囊激肽（CCK）被认为是一种短期饱腹感因子，参与减少和终止摄食，也启动行为饱腹感序列，因此参与调节食物摄入量（Cummings and Overduin，2007）。在缺氧条件下，我们还观察到 *CCK* 表达的上调。Bailey et al.（2000）提出 *CCK* 表达的变化是缺氧期间高海拔厌食的原因。结果表明，在洛氏鲅中，*CCK* 表达上调导致食物摄入量减少。还应该注意的是，鱼只处理了10d，因此需要更多的时间点结合持续和昼夜低氧处理来充分阐明结果。

此外，缺氧应激会诱导和抑制鱼类中的一系列基因表达，并会破坏生物钟相关基因的表达模式（Connor and Gracey，2011）。已经发现，生物钟基因有节奏的基因表达模式在各种条件下受到调节，包括食物信号、水温和外源性毒素（Vera et al.，2013；Yin et al.，2020）。除了葡萄糖代谢基因（*Glut1* 和 *CCK*）外，研究的8个生物钟基因在持续缺氧和昼夜缺氧条件下表现出显著差异。以前在可比较的低氧处理期间检测到生物钟基因的表达差异（Betancor et al.，2020；Prokkola et al.，2015），但迄今为止，很少有研究调查鱼类中昼夜循环低氧导致的生物钟基因表达模式。基因组数据库显示，鱼类的生物钟基因数量是不同的。斑马鱼和日本河鲀各有三个生物钟基因，而日本青鳉（*Oryzias latipes*）、三刺棘鱼（*Gasterosteus aculeatus*）和绿河鲀（*Tetraodon nigroviridis*）有两个生物钟基因（Wang，2008）。Clock蛋白参与小鼠肝脏中许多昼夜节律输出基因的转录调控，并且似乎还参与各种生理功能，如细胞周期、脂质代谢、免疫功能和外周组织中的蛋白水解（Oishi et al.，2003）。与对照相比，肝脏 *Clock a* 基因在昼夜低氧处理条件下表现出上调，在持续低氧处理条件下表现出轻微下调，表明节律振荡被不同的低氧模式破坏。同时，昼夜低氧胁迫条件下 *Clock a* 基因的上调以及 *Cry1a* 基因的下调与脊椎动物生物钟的典型模型一致（Cahill，2002）。在洛氏鲅的肝脏中，*Cry1a* 和 *Cry1b* 基因的表达在持续低氧处理和昼夜低氧处理期间有所不同。在持续低氧处理和昼夜低氧处理期间，*Cry1a* 基因的表达显著下调，而 *Cry1b* 基因表达显著上调。*Cry1a* 和 *Cry1b* 对负反馈回路有不同的影响，负反馈回路会产生基因表达的昼夜节律振荡（Kobayashi et al.，2000）。肝脏特异性的Cry1耗竭已被证明会刺激肝细胞从头产生葡萄糖，从而导致血糖浓度升高（Zhang et al.，2010）。在大西洋鳕快骨骼肌中也发现了旁系同源物之间 *Cry1a* 和 *Cry1b* 基因表达模式的明显差异，其中Cry1a在18个组织的13个中被检测到，而Cry1b仅存在于 11 个组织中（Lazado et al.，2014）。同时，我们的结果表明 *Per1a* 基因表达最初被诱导，然后在持续低氧处理和昼夜低氧处理下被抑制。在三刺棘鱼中发现 *Per1a* 基因表达在低氧 10.5h 后增加，但在低氧 24h 后降低，作者假设 *Per1a* 的启动子可能对 HIF-1α 敏感（Prokkola et al.，2015）。有报道发现，小鼠持续低氧会增加大脑中 Per1 和 Clock 的蛋白质水平（Chilov et al.，2002）。Per2 被认为是一种控制细胞增殖和凋亡的成分（Wang et al.，2016）。昼夜节律 *Per2* 表达的中断对核心时钟动力学及其在脂质代谢方面的其他功能（Grimaldi et al.，2010）和氨基酸转运蛋白的 mRNA 丰度（Spanagel et al.，2005）有影响。我们的结果表明洛氏鳅 *Per1* 对缺氧（尤其是昼夜低氧处理）比 *Per2* 更敏感。此外，*Per2* 转录抑制可能与 HIF-1α 有关，HIF-1α 与 Clock 的 *Per* 基因启动子区域中的相同序列竞争性结合，使得 *Per* 转录的昼夜节律在低氧条件下受到抑制（Egg et al.，2013；Pelster and Egg，2015；Prokkola and Nikinmaa，2018）。结构上的相似性表明哺乳动物 Timeless 可能作为生物钟内的分子成分发挥作用，通过与 *Per1* 结合，导致 *Per1* 表达的抑制（Koike et al.，1998）。Timeless 对 *Per1* 的抑制作用已在果蝇中得到证实（Koike et al.，1998），该研究在结构上与我们目前的研究相似。我们的结果表明，除了第 8 天的昼夜低氧处理之外，*Timeless* 在持续低氧处理和昼夜低氧处理期间被下调。

HIF-1 是一种由低氧激活的异二聚体转录因子，它迅速积累并转移到细胞核，在那里与 HIF-1β 和 p300/CBP 二聚体形成（O'Connell et al.，2020）。短期和长期的低氧暴露已被证明会导致黄花鱼中 *HIF-1α* mRNA 表达显著增加，这被视为黄花鱼适应慢性低氧的重要组成部分（Rahman and Thomas，2007）。已经证明 HIF-1α 负责使生物钟适应低氧张力（Pelster and Egg，2018）。动物中 *Per1*、*Per2* 和 *Cry2* 的表达水平在 HIF-1α 敲除后严重降低，证明了 HIF-1α 信号与生物钟之间的相互作用（Adamovich et al.，2016）。我们的结果表明，昼夜低氧处理和持续低氧处理暴露均诱导了洛氏鳅中 *HIF-1α* 基因的表达。随着低氧暴露时间的增加，*HIF-1α* 的表达呈先下降后上升趋势，这与 *Cry1a* 转录本的表达模式几乎完全相反。这证实了高纬度鱼类洛氏鳅在缺氧条件下 HIF-1α 信号与生物钟的相互作用，但具体机制尚不清楚。

第五章　洛氏鲹昼夜低氧和持续低氧处理后心脏和脑的响应机制

目前关于鱼类低氧研究主要集中在肝脏代谢与低氧暴露等内容上，分析低氧条件下鱼的心脏和大脑应答机制的研究较为匮乏。心脏作为血液运输的主要器官，是推动血流为身体提供氧气和营养的场所（Hall，2015）。急性低氧可导致心脏被优先灌注以保护机体，但持续低氧会对心脏产生不利影响，因为心脏代偿和保护机制仅能维持在有限的时间内（Martínez et al.，2006；Kolár and Oštádal，2004）。因此，迫切需要了解水生动物的心脏将如何应对持续的以及循环的低氧环境。与其他组织相比，大脑对能量和氧气有着积极的需求。然而目前，人们对于低氧暴露对鱼类脑组织学的影响知之甚少。在鱼类中，慢性低氧导致心室流出道变小，心脏腔隙减少，心肌细胞密度增加，然而，慢性持续低氧可能导致心肌细胞核密度增加，表明心肌细胞和/或核增生（Marques et al.，2008）。此外，通过比较发现，昼夜低氧和持续低氧暴露模式可能不会影响鱼类大脑或心脏中代谢酶的活动，但间歇性低氧会增加大脑质量（Borowiec et al.，2015）。低氧应激条件下，鱼类心脏的免疫过程、糖酵解途径和离子转运发生了显著变化，这些变化可能有助于在低氧期间维持细胞能量平衡。而鱼类也会通过抗氧化酶活的活性增加，应对不良的氧气环境。

有研究表明尽管大脑仅占生物体重的2%，但它消耗了总氧气的20%（Yu and Li，2011；Ekambaram et al.，2017）。哺乳动物的大脑对氧气缺乏非常敏感，随着氧气可用性降低和有氧ATP产生下降，ATP依赖性离子泵功能受损，导致神经元膜去极化、电压敏感离子通道激活、钙离子过度流入、神经元过度兴奋和兴奋性毒性，以及细胞死亡（Farhat et al.，2021）。也有研究认为，低氧会导致脑血流量增加（Martínez et al.，2006；Lawley et al.，2017；Gibbons et al.，2019），而这种生理压力会改变活性氧（ROS）的含量，导致氧化应激和细胞功能丧失，并最终导致细胞凋亡或坏死（Halliwell and Gutteridge，1984）。鱼的大脑在适应环境变化方面也发挥着重要作用。有研究证实在葡萄糖供应有限的情况下，硬骨鱼的大脑会利用其他燃料（如乳酸），但是关于在鱼脑中使用脂质和氨基酸作为能量来源的信息很少（Soengas and Aldegunde，2002）。鲤器官中抗氧化防御的反应体现在大脑中一些抗氧化酶在低氧期间的增加，为了减少生物体内的活性氧，鲤激活了抗氧化酶防御系统（Lushchak et al.，2005）。但也有研究报道认为持续低氧可显著降低虾虎鱼（*Perccottus glenii*）的大脑中抗氧化酶活性（Lushchak and Bagnyukova，2007）。

一、昼夜低氧和持续低氧处理后心脏的组织学变化

光镜观察的心脏，固定后通过乙醇-二甲苯系列脱水，然后真空包埋在石蜡中。将样品切片（7μm）并固定在涂有聚赖氨酸的载玻片上，切片采用苏木精和伊红（H&E）染色，并在显微镜下成像。扫描电镜的心脏，采用戊二醛固定且干燥后，固定在铜架上，采用 Hitachi E-1010 仪器喷金，后采用 Hitachi S 2700 仪器进行观察。透射电镜样品通过戊二醛的固定后，采用锇酸固定，后采用丙酮脱水后，醋酸铅染色，用 Reichert-Jung Ultracut E 仪器进行切片后，用 JEM-1200 EX II 进行观察。

（一）心脏的组织学变化

观察发现，经过持续低氧和昼夜低氧处理，洛氏鱥的心肌细胞呈现明显的变化。如彩图 7 所示，在持续低氧处理后，洛氏鱥的心肌细胞核的数量显著下降（$P<0.05$），在处理 21 和 28d 明显低于对照组，在持续低氧 28d 为最低值。同样，经过昼夜低氧处理后，洛氏鱥的心肌细胞核的数量也呈现显著下降（$P<0.05$），并在处理 28d 为最低值。

（二）综合分析

心脏是维持血液运输的重要器官，大量的报道证实了心脏经过有毒物质胁迫后的形态变化，但是关于持续低氧和昼夜低氧胁迫后的比较研究尚无。有报道表明，当暴露于慢性持续低氧时，斑马鱼和慈鲷的心脏会显著缩小心室流出道并减少中央心室腔内的腔隙（Marques et al.，2008）。然而，据我们所知，在鱼类中尚未报告持续或循环低氧引起的心室改变，我们的发现值得进一步调查。在本研究中，该鱼的心室流出道和腔隙在持续低氧后明显增加了。此外，使用扫描电子显微镜证实了光镜观察下心脏中的这些变化（图 5-1）。在鲢的研究中发现，低氧会导致心肌细胞间隙增大，心肌纤维出现紊乱、肿胀甚至破裂，认为其心肌纤维在低氧处理过程中出现了严重的损伤（Li et al.，2021）。一些研究认为，动物主要通过剩余心肌细胞的肥大来补偿心肌细胞的这种损失（Ostadal and Kolar，2007），虽然可能会发生心肌细胞增殖引起的心肌再生，但扩展程度十分有限（Beltrami et al.，2001）。我们的结果证实，在这两种低氧条件下，心脏的响应机制是不同的，洛氏鱥可能通过在持续低氧期间发生心肌细胞的肥大来适应低氧。

第五章 洛氏鲅昼夜低氧和持续低氧处理后心脏和脑的响应机制

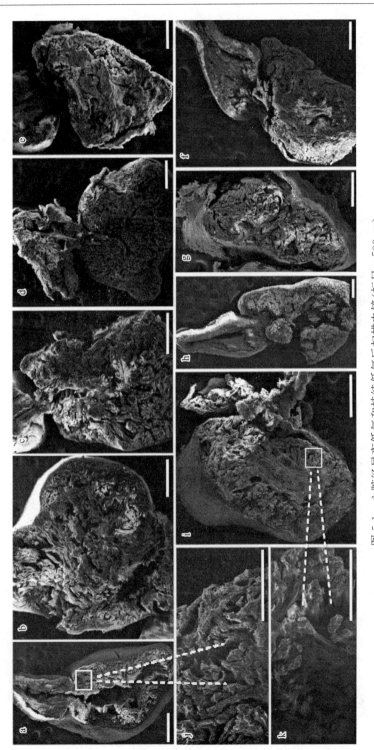

图 5-1 心脏经昼夜低氧和持续低氧后扫描电镜(标尺=500μm)

a. 对照组 b. 昼夜低氧 7d c. 昼夜低氧 14d d. 昼夜低氧 21d e. 昼夜低氧 28d f. 持续低氧 7d g. 持续低氧 14d h. 持续低氧 21d i. 持续低氧 28d j. 对照组放大图片 k. 昼夜低氧 28d 放大图片

二、昼夜低氧和持续低氧处理后脑的组织学变化

(一) 脑的组织学变化

1. 持续低氧处理后脑组织学变化（光镜观察） 低氧处理后，我们观察了洛氏鱥的中脑各层变化，主要包括硬脑膜层（dura），表面纤维层（stratum fibrosum et griseum superficiale）、视层（stratum opticum）、边缘纤维层（stratum fibrosum marginale）和围脑室层（Stratum periventricular）（彩图 8~彩图 10）。我们的结果显示，持续低氧 7d 后硬脑膜层的比例显著增加。

2. 昼夜低氧处理后脑组织学变化（光镜观察） 见彩图 8~彩图 10，昼夜低氧 21d 后，洛氏鱥的顶盖中央白质层（stratum album centrale）比例显著增加。与其他组相比，昼夜低氧 28d 后，表面纤维层、视层的比例显著增加，而昼夜低氧 28d，边缘纤维层减少。此外，昼夜低氧 14d 的围脑室层比例显著降低。

3. 持续和昼夜低氧处理后脑组织变化扫描电镜（SEM）和透射电镜（TEM）观察 TEM 结果显示持续低氧后脑细胞中存在有髓轴突，在对照中也可见，在昼夜循环低氧组未观察到有髓轴突（图 5-2~图 5-4）。

(二) 综合分析

关于低氧胁迫后鱼类局部脑部组织学观察的详细报告相对较少，其中大部分是关于饮食处理和毒性暴露的研究。在本研究中，持续的和循环的低氧都会导致中脑层的相对厚度发生变化。与我们的结果类似，虹鳟在硫酸铜暴露后也显示出中脑的纤维和灰层减少（Al-Bairuty et al.，2013）。在对雏鸡的研究中，低氧导致胚胎大脑中脑视顶盖（stratum griseum central，SGC）层的死亡，这些层中的每一层对低氧事件都有不同的敏感性变化，这表明中脑的每一层对低氧的反应可能不同（Pozo Devoto et al.，2006）。在用毒死蜱处理印度囊鳃鲇（*Heteropneustes fossilis*）后，发现边缘层（stratum marginale，SM）和视层轻微脱离，SPV（stratum periventricular）区颗粒细胞变性和迁移，单个核细胞海绵状组织增生和坏死（Mishra et al.，2020）。

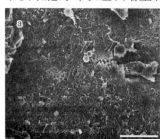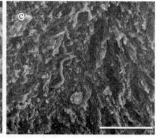

第五章 洛氏鲹昼夜低氧和持续低氧处理后心脏和脑的响应机制

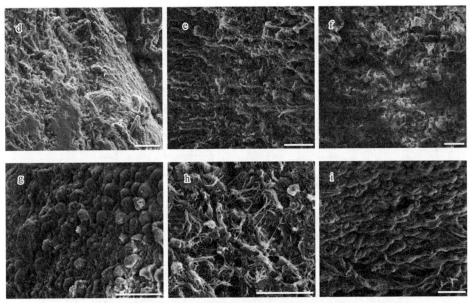

图 5-2 脑经持续和昼夜低氧后的扫描电镜图（标尺=20μm）
a. 对照组　b. 昼夜低氧 7d　c. 昼夜低氧 14d　d. 昼夜低氧 21d　e. 昼夜低氧 28d
f. 持续低氧 7d　g. 持续低氧 14d　h. 持续低氧 21d　i. 持续低氧 28d

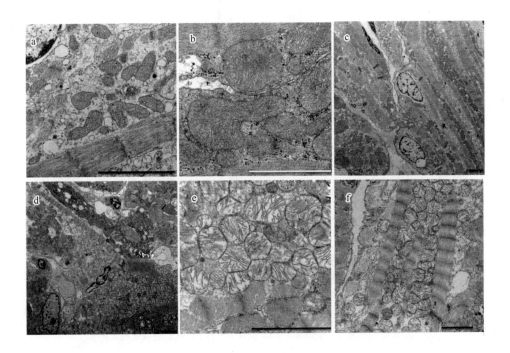

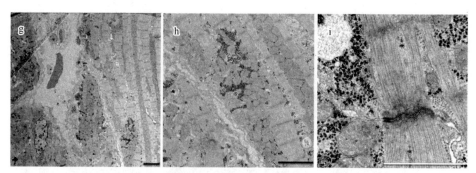

图 5-3 心脏经持续和昼夜低氧后的透射电镜图（黑色标尺＝500nm，白色标尺＝2μm）
a、b 和 c. 对照组　d、e 和 f. 持续低氧 28d　g、h 和 i. 昼夜低氧 28d

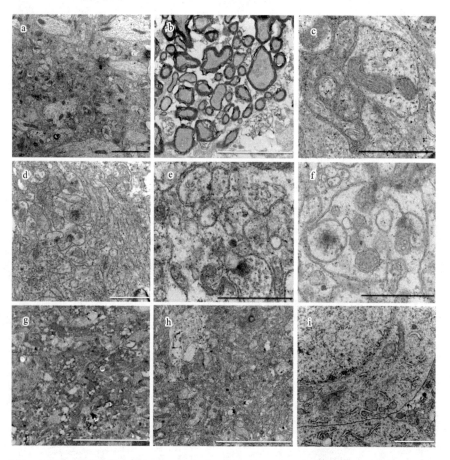

图 5-4 脑经持续和昼夜低氧后的透射电镜图（黑色的标尺为 500nm，比例尺＝2μm）
a、b 和 c. 对照组　d、e 和 f. 持续低氧 28d　g、h 和 i. 昼夜低氧 28d

三、昼夜低氧和持续低氧处理后心脏和脑中抗氧化酶活性的变化

(一)昼夜低氧和持续低氧处理后心脏中抗氧化酶活性的变化

洛氏鱥心脏中抗氧化酶在低氧处理后的变化情况见图5-5。心脏中超氧化物歧化酶的酶活性在持续低氧组呈现先下降后上升的趋势,持续低氧处理14d后,心脏中SOD酶活性显著降低($P<0.05$),但是其他各组变化不显著($P>0.05$)。过氧化氢酶在持续低氧和昼夜低氧处理组中都没有显著性变化($P>0.05$),仅在昼夜低氧处理后21d,酶活性显著升高($P<0.05$)。在整个实验中,谷胱甘肽过氧化物酶在昼夜低氧28d时显著增加,但是在其他组没有显著性变化。丙二醛在持续低氧胁迫时,最高值出现在持续低氧28d,但是与对照组相比,各组没有显著性变化($P>0.05$)。在昼夜低氧处理时,最高值出现在21d,与各组相比没有显著变化($P>0.05$)。

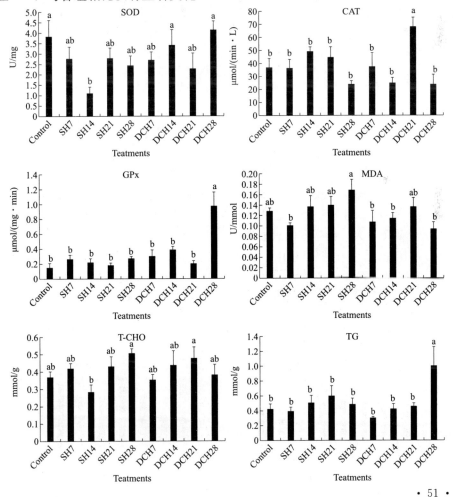

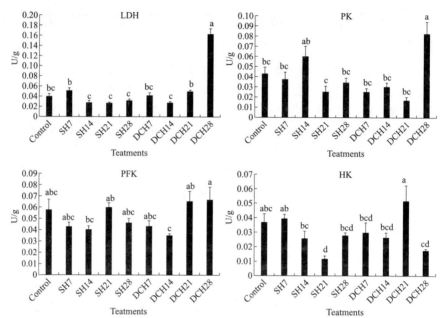

图 5-5　洛氏鱥持续低氧和昼夜低氧处理后心脏中的酶活变化

注：图中标有不同字母表示差异显著（$P<0.05$），否则差异不显著。

（二）昼夜低氧和持续低氧处理后脑中抗氧化酶活性的变化

洛氏鱥脑中抗氧化酶在低氧处理后的变化情况见图 5-6。经过持续低氧处理 28d，脑中超氧化物歧化酶的酶活性没有显著性变化（$P>0.05$），但是在昼夜低氧胁迫 28d，与对照组相比脑中 SOD 酶活性显著上升（$P<0.05$），并且在 28d 达到最高值，其他各组与对照组不存在显著差异。过氧化氢酶的酶活性在持续低氧时持续上升，处理后 21d 显著高于对照组（$P<0.05$），且 CAT 活性在 21d 达到了最大，在昼夜低氧处理后的 14d 和 28d 时显著高于对照组（$P<0.05$），其他各组与对照组相比不存在显著差异。谷胱甘肽过氧化物酶在持续低氧 7d 后显著增加（$P<0.05$），达到最大值，在昼夜低氧 21d 和 28d 后降低（$P<0.05$）。丙二醛的最高值出现在昼夜低氧 28d，与对照组相比，其他各组没有显著性变化（$P>0.05$）。

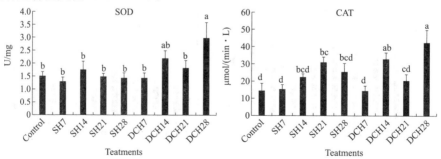

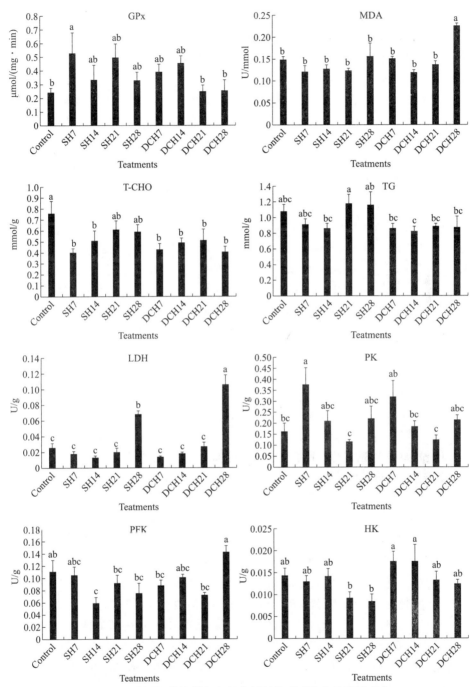

图 5-6 洛氏鱥持续低氧和昼夜低氧处理后脑中的酶活变化

注：图中标有不同字母表示差异显著（$P<0.05$），否则差异不显著。

（三）综合分析

一种普遍存在的机体对低氧变化的反应是活性氧（ROS）的过度产生（Yang et al., 2017，Wang et al., 2021b）。此外，响应升高的 ROS 水平诱导的抗氧化酶机制（如过氧化氢酶、超氧化物歧化酶）可以作为抗氧化剂（Scanalios, 2005，Vutukuru et al., 2006）。有趣的是，本研究实验结果表明经过长期持续和昼夜低氧处理后，洛氏鱥的大脑没有遭受氧化损伤。与若鲉杜父鱼（*Scorpaenichthys marmoratus*）分析的结果类似，大脑低氧后未观察到任何反应（Lau et al., 2019）。Wang et al. (2021) 实验结果表明在 10d 的低氧处理时，持续和昼夜低氧会降低洛氏鱥肝脏中的 SOD 酶活性。一些研究表明，在低氧期间，血液会优先分配到需氧量高的关键器官，如大脑和心脏当中（Perry et al., 2009）。此外有报道认为，似鳊驼背脂鲤大脑的低氧抗氧化反应与其他组织无关（Johannsson et al., 2018）。本研究中 CAT 酶活性显著增加，表明 CAT 在持续和昼夜循环低氧暴露中发挥抗氧化作用。当通过低氧产生 ROS 时，CAT 和 GPx 进一步将 H_2O_2 转化为 H_2O 和 O_2，保护细胞免受氧化损伤（Cao et al., 2012）。可能还需要注意的是，虽然在暴露于两种低氧模式期间，心脏和大脑中 SOD 和 CAT 的活性发生显著变化，但 MDA 含量并没有减少，这可能与之暴露于低氧后抗氧化能力的降低有关（Johannsson et al., 2018）。

四、昼夜低氧和持续低氧处理后脑和心脏中糖脂代谢酶指标的变化

（一）昼夜低氧和持续低氧处理后心脏中糖代谢酶活性的变化

洛氏鱥心脏中糖代谢相关的葡萄糖含量，以及 PFK、LDH、HK 和 PK 的酶活性变化情况见图 5-5。HK 活性在持续低氧处理时呈现下降趋势，在 21d 时显著降低，且达到最低值。HK 活性在昼夜低氧 21d 时值最高，28d 时值最低。PFK 活性在持续低氧时，各组与对照组没有显著性差异（$P>0.05$），在昼夜低氧处理后 14d 时值最低，后上升，在处理 28d 时值最高。洛氏鱥的 LDH 活性在持续低氧处理后，各组与对照组相比没有显著性差异（$P>0.05$），但是在昼夜低氧处理 28d 时，LDH 活性比对照组显著上升（$P<0.05$），且达到最大值。PK 活性在持续低氧处理 21d 时达到最低值，但是与对照组相比差异不显著（$P>0.05$）。PK 活性在昼夜低氧处理 28d 时比对照组和其他各组显著升高（$P<0.05$），达到最大值，其他各组与对照组相比差异不显著（$P>0.05$）。

（二）昼夜低氧和持续低氧处理后脑中糖代谢酶活性的变化

洛氏鱥脑中糖代谢相关的葡萄糖含量，以及 PFK、LDH、HK 和 PK 的

酶活性变化情况见图5-6。HK活性在持续低氧各组间没有显著性差异，在昼夜低氧处理时的14d时值最大。PFK活性在持续低氧处理第14d时比对照组显著降低（$P<0.05$），但是其他各组与对照组相比无显著性差异（$P>0.05$）。PFK活性在昼夜低氧处理时，处理组与对照组之间没有显著性差异（$P>0.05$），在28d时活性最高。LDH活性在持续低氧处理的28d显著高于其他各组（$P<0.05$），昼夜低氧处理28d时显著高于其他各组（$P<0.05$），其他组与对照组和各组之间无显著差异（$P>0.05$）。PK活性在持续低氧组7d时显著高于对照组（$P<0.05$），在其他处理时间与对照组和各组之间无显著差异（$P>0.05$）。

（三）昼夜低氧和持续低氧处理后脑中甘油三酯和胆固醇的含量变化

洛氏鱥脂代谢甘油三酯和胆固醇的含量变化情况见图5-6。与对照组相比，甘油三酯在持续低氧时呈现先下降后上升的趋势，在持续低氧21d时值达到最大，但是在昼夜低氧时没有明显变化（$P>0.05$）。胆固醇在持续低氧7d和14d时显著降低（$P<0.05$），在持续低氧的其他各组变化不显著（$P>0.05$），但是在昼夜低氧处理后的各组显著降低（$P<0.05$）。

（四）昼夜低氧和持续低氧处理后心脏中甘油三酯和胆固醇的含量变化

洛氏鱥经过持续低氧处理后，如图5-5所示，心脏中CHO含量在14d时达到最低值，但是各组没有显著性差异（$P>0.05$），经昼夜低氧处理后，CHO含量在7d时达到最低值，各组与对照组相比没有显著性差异（$P>0.05$）。洛氏鱥经过持续低氧处理后，TG含量没有显著性变化（$P>0.05$），但是经过昼夜低氧处理后，TG含量在28d时显著高于对照组和其他各组，且达到最大值（$P<0.05$）。

（五）综合分析

耐低氧生物使用代谢抑制作为应对氧含量降低的关键策略，否则低氧对大多数动物产生有害影响（Bickler and Buck，2007；Richars，2010）。我们之前的研究已经证实洛氏鱥是一种耐低氧鱼（Yang et al.，2021）。之前的研究已经表明，胆固醇（Na^+-K^+-ATP酶调节剂）和二十二碳六烯酸百分比的变化可能与降低代谢率有关（Harayama and Riezman，2018；Farhat et al.，2020）。在本研究中，持续低氧处理28d组心脏胆固醇水平升高。与我们的研究类似，大鼠的慢性低氧表明低氧会导致心脏、肌肉、肾脏中胆固醇丰度发生巨大变化，并且在大脑和肝脏中降低（Farhat et al.，2020）。此外，我们的数据强烈表明，持续低氧组中的胆固醇升高，但昼夜循环低氧组没有显著变化，这一发现表明鱼类响应持续低氧的能量储存和利用的能力明显高于昼夜循环低氧组的鱼。本研究证明了低氧后洛氏鱥心脏和大脑中甘油三酯的显著增加。在对另一种冷水鱼细鳞大麻哈鱼（*Oncorhyncus gorbuscha*）的研究中发现，甘油三酯代谢途径是鱼心脏中比较活跃的代谢途径之一，心脏比肝脏、骨骼肌或大脑更容易从血清中吸收和保留脂肪酸。

除了脂质代谢，葡萄糖代谢是鱼类适应低氧的另一种方式，通过提供能量（Li et al., 2018；Sun et al., 2021）。在对大黄鱼进行低氧条件下的心脏组织分析时，心脏的糖酵解途径和离子转运发生显著改变，作者认为这可能有助于在低氧期间维持细胞能量平衡（Mu et al., 2020）。LDH 和 PK 先前已被证明在暴露于环境压力的鱼中活性受到影响（Pierce and Crawfor, 1997）。我们的结果同样显示了昼夜低氧处理 28d 组心脏和大脑的 LDH 酶活性显著增加。有研究观察到耐低氧杜父鱼的脑 LDH 水平高于低氧敏感的物种，这表明大脑主要依赖糖原和葡萄糖为耐低氧鱼类提供能量（Mandic et al., 2012）。同时，本研究强调心脏和脑在持续低氧后 LDH 酶活性无显著差异，但 PK 酶活性增加。为了维持代谢能量平衡并防止低氧诱导的细胞死亡，耐低氧动物会下调 ATP 消耗途径，并通过增加糖酵解将 ATP 生成过程转为底物磷酸化（Mandic et al., 2012）。我们认为洛氏鲅对持续低氧的适应主要是通过增加 PK 酶的活性。此外，本研究中，PFK 酶活性在循环低氧 28d 组中显著增加，表明昼夜循环低氧模式下的鱼类使用葡萄糖代谢来防止间歇性低氧诱导的细胞死亡。研究结果表明，洛氏鲅在持续低氧期间主要利用 LDH 和 PK，但在循环低氧（尤其是长期应激）期间更充分利用心脏和大脑中的葡萄糖和脂质代谢。

五、昼夜低氧和持续低氧处理后脑和心脏中低氧信号传导基因表达的变化

（一）昼夜低氧和持续低氧处理后心脏中低氧信号传导基因表达的变化

洛氏鲅心脏低氧相关信号通路基因 *HIF*、*VHL*、*FIH* 和 *P53* 的基因表达见图 5-7。在心脏中，经持续低氧和昼夜低氧处理后，*HIF-1α* 相对表达水平在昼夜低氧组 28d 时显著高于对照组和其他各组（$P<0.05$），其他组与对照组和各组之间无显著性差异（$P>0.05$）。*HIF-2α* 相对表达水平在持续低氧时各组没有显著性差异（$P>0.05$），在昼夜低氧 7d 时显著高于对照组（$P<0.05$），在 28d 时显著高于对照组和其他各组（$P<0.05$），且在 28d 时达到最大值。*HIF-3α* 相对表达水平在持续低氧处理的组间没有显著差异，在持续低氧 14d 时值最大，并且与对照组相比差异显著，在昼夜低氧 28d 时显著高于对照组和各组（$P<0.05$），且为最高值。*VHL* 相对表达水平在持续低氧仅在 14d 表现出明显差异，在昼夜低氧 14d 和 28d 时表达量明显高于对照组和其他处理组（$P<0.05$），但是在其他组与对照组和各组之间没有显著性差异。

FIH 相对表达水平在持续低氧 14d 时显著增加（$P<0.05$），但是其他组没有显著性变化。FIH 相对表达水平在昼夜低氧 28d 时显著增加，显著高于对照组和其他组（$P<0.05$），其他组无显著性变化。P53 相对表达水平仅在昼夜低氧 28d 时比对照组和各组显著升高（$P<0.05$），且为最高值，其他组没有显著性变化（$P>0.05$）。

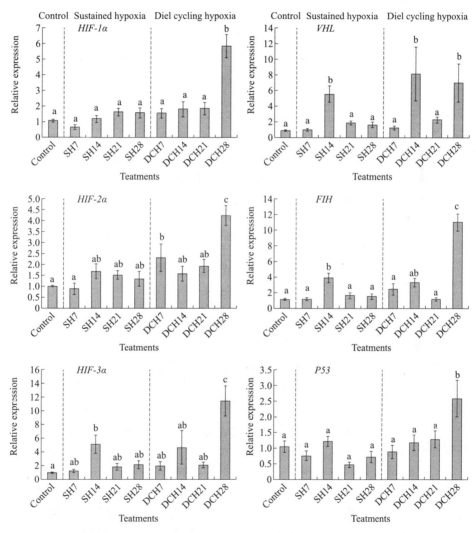

图 5-7　洛氏鱥的心脏持续和昼夜低氧处理后，低氧转导信号相关基因的表达模式
注：图中标有不同字母表示差异显著（$P<0.05$），否则差异不显著。

（二）昼夜低氧和持续低氧处理后脑中低氧信号传导基因表达的变化

洛氏鱥脑的低氧相关信号通路基因 HIF、VHL、FIH 和 P53 的基因表达见图 5-8。在脑中，$HIF\text{-}1\alpha$ 相对表达水平在持续低氧处理 28d 时达到最高，显著高于处理 14d 和 21d 组（$P<0.05$），在昼夜低氧处理时，没有明显变化（$P>0.05$）。$HIF\text{-}2\alpha$ 相对表达水平在持续低氧 28d 时达到最高值，显著高于处理 7d（$P<0.05$）。在昼夜低氧处理时，各组之间没有显著性差异（$P>0.05$）。$HIF\text{-}3\alpha$ 相对表达水平在持续低氧处理的组间没有显著差异，在持续低氧 28d 时值最大，在昼夜低氧 7d 和 14d 时显著低于对照组（$P<0.05$）。在脑中，VHL 相对表达量在持续低氧 28d 显著升高，高于对照组和其他各组（$P<0.05$），为最高值。在昼夜低氧处理时，各组之间没有显著性差异，与对照组也无显著性差异。FIH 基因表达量在经过持续低氧处理后，在 7d 时达到最高值，显著高于对照组、处理 14d 和 21d 组（$P<0.05$）。FIH 基因表达量在昼夜低氧处理时呈现先下降后上升的趋势，在处理 28d 时达到最高。昼夜低氧处理的 FIH 在 14d 和 21d 显著低于 28d（$P<0.05$）。$P53$ 基因的相对表达水平仅在持续低氧处理后显著下调，尤其是处理 21d 显著低于处理 7d 和 28d（$P<0.05$）。$P53$ 基因表达量在昼夜低氧处理后的 21d 达到最低值，显著低于对照组和其他各组（$P<0.05$）。

（三）综合分析

作为低氧诱导因子（HIF）的天冬酰胺羟化酶，低氧诱导因子抑制因子 FIH 通过阻断 HIF 与转录共激活因子 CREB 结合蛋白（CBP）和 p300 的结合，来抑制低氧诱导基因的转录激活（Li et al., 2008）。近来，FIH1 被鉴定为一种分子氧依赖性双加氧酶，它可减弱 HIF 的转录活性，并且还与 HIF 依赖性和非依赖性的低氧反应有关（Schodel et al., 2010）。在正常氧气条件下，FIH 羟基化 $HIF\text{-}1\alpha$ 的 CAD 结构域上的天冬酰胺残基，然后阻断 p300 和 CBP 等转录辅激活因子与 $HIF\text{-}1\alpha$ 的 CAD 结合，从而抑制 $HIF\text{-}1\alpha$ 介导的基因转录（Mahon et al., 2001；Lando et al., 2002）。然而，在低氧条件下，羟基化被消除，允许 $HIF\text{-}1\alpha$ 募集 CBP/p300，从而促进目标基因的表达（Freedman et al., 2002）。FIH-1 还与一系列其他底物中的羟基化事件有关，包括 Notch 途径（Wilkins et al., 2009）和端锚聚合酶蛋白（Cockman et al., 2009）。此外，Von Hippel-Lindau（VHL）是一个典型的肿瘤抑制基因，其失活与大量的器官肿瘤发生有关（Kaelin et al., 2007）。VHL 的功能主要是作为多蛋白 E3 泛素连接酶复合物的底物识别亚基发挥作用，该复合物靶向脯氨酸羟基化低氧诱导因子（HIF）$1/2\alpha$，在正常氧气条件下进行蛋白酶体降解（Kaelin et al., 2007；Maher and Kaelin, 1997）。因此，pVHL 是低氧信号传导的关键调节因子。目前，关于不同低氧条件下的低氧信号传导调节因子的表达模式研究较为缺乏。

第五章 洛氏鲅昼夜低氧和持续低氧处理后心脏和脑的响应机制

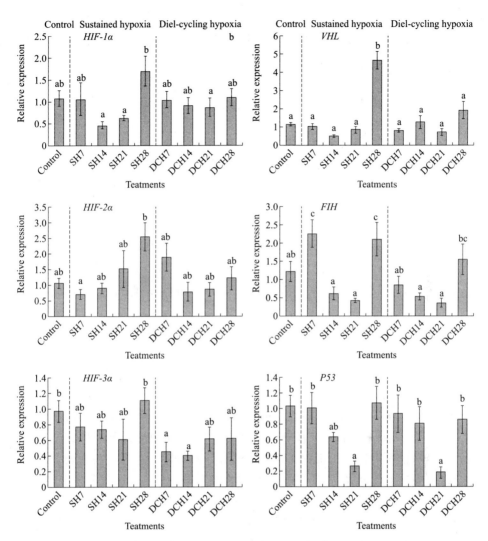

图 5-8 洛氏鲅的脑持续和昼夜低氧处理后，低氧转导信号相关基因的表达模式
注：图中标有不同字母表示差异显著（$P<0.05$），否则差异不显著。

通过基因表达的研究来表征鱼类在暴露于低氧应激源时的响应机制是直观且十分必要的。目前普遍认为，HIF-1α 调节急性低氧反应，HIF-2α 调节慢性低氧反应，而 HIF-3α 抑制其他两种同工型的活性（Holmquist-Mengelbier et al.，2006，Zhang et al.，2014）。在低氧时，HIF-1α 在细胞中积累并与 HIF-1α 形成二聚体，从而促进参与血管生成和糖酵解等各种过程的基因的表达（Neufeld et al.，1999；Schofield and Ratcliffe，2004）。在分子水平上，低氧诱导因子（HIF）可以通过调节几个基因级联的转录来感知细胞内氧气的减少并激活细胞反应途径（Zhu et al.，2013）。在对低氧处理的胡鲇（*Clarias batrachus*）的研究中，急性暴露会诱导大脑 *HIF-1α* 和 *HIF-2α* 表达，但是，长期暴露期间表达没有显著差异（Mohindra et al.，2013）。本研究中，与对照组相比，洛氏鲅脑中 *HIF-1α* 和 *HIF-2α* 的表达水平升高，并且在持续低氧处理 28d 组最高（但在昼夜低氧处理 28d 组中并没有升高），表明 HIF 在持续长期低氧条件下发挥作用。此外，在心脏中，HIF-1α 和 HIF-2α 在持续低氧处理 7d 时减少，在洛氏鲅的其他组中增加。这与肩章鲨（*Hemiscyllium ocellatum*）的研究类似，即在反复的低氧暴露中，以组织特异性方式激活 HIF-1α 通路和下游保护通路的转录（Rytkönen et al.，2012）。在黄尾平口石首鱼（*Leiostomus xanthurus*）中发现，与正常氧气对照组相比，暴露于持续和昼夜循环低氧 3d 后，肌肉组织中的 HIF-1α 蛋白增加（$P<0.005$），而在所有处理中，昼夜循环低氧产生的 HIF-1α 蛋白浓度高于持续低氧，表明氧气的波动诱导了 *HIF-1α* 的基因表达（Smith et al.，2012）。

许多研究已经证明，随着长期驯化，HIF-1α 丰度的降低可能是由于氧气浓度的改变或对氧气敏感性的变化（如脯氨酰羟化酶活性或氧气动力学的变化）所致（Wenger，2002；Kopp et al.，2011）。上述数据表明，鱼类中 *HIF-α* 亚基的差异表达具有低氧处理时间和组织的依赖性，并且可能在对低氧的适应性反应中发挥特殊作用。除了 *HIF* 基因（*HIF-1α*、*HIF-2α*、*HIF-3α*，*HIF-1β*），低氧信号还包括 HIF 脯氨酸羟化酶基因（*PHD-1*、*PHD-2* 和 *PHD-3*）、天冬氨酸羟化酶基因（*FIH-1*）和 Von Hippel-Lindau 基因（*VHL*）（Zhang et al.，2017）。在常氧条件下，PHD 羟基化 HIF-α 的脯氨酸残基并将其与 VHL 结合，羟基化的 HIF-α 泛素化（Wang et al.，2021b）。然而，在低氧条件下，由于缺乏氧分子作为底物，PHD 不能羟基化 HIF-α，并引起一系列生理生化反应（Xiao，2015）。许多研究证实，在鱼类急性低氧应激期间，*VHL* 和 *HIF* 的表达呈现出一致的变化（Zhang et al.，2017；Wang et al.，2021b），但是鱼类长期低氧中关于 *VHL* 表达的研究很少。洛氏鲅中的 VHL 在心脏中下调，但在大脑中的持续低氧 28d 和昼夜低氧 28d 上调，表明它在长期低氧处理的大脑中起关键作用。一些研究表明 *VHL* 显著表达，是由于低氧后底物的缺乏，阻止它们羟基化 HIF 的特定脯氨酸位点，最终 VHL 不能被泛素化（Wang et al.，2021b）。此外，研究证实，*VEGF* 的转录表达在不同组织中存在特异性。例如，在黄尾平口石首鱼的肌肉中，低氧反应通路的调节方式与肝脏不同，该研究认为，随着低氧发生，*VEGF* 的转录物数量减少，与肌肉中 HIF 的不稳定有一定关系（Smith et al.，2012）。在低氧信号转导中，FIH-1 在正常氧气条件下羟基化 HIF-α 的特定天冬酰胺残基，抑制 HIF 与 CBP/p300 复合物结合，从而增强转录 HIF 通路的活性。在低氧条件下，HIF-1α 羟基化能力降低，HIF-1α 可以与转录辅因子结合，促进下游靶基因的转录（Xiao，2015；Zhang et al.，2016；Okamura et al.，2018）。在低氧条件下的团头鲂 *FIH* 表达的研究中发现，FIH 进行的 HIF-1α 羟基化过程减弱/消失，加速了低氧条件下 HIF-1α 的积累，并且大量的 HIF-1α 诱导了 *FIH* 表达（Zhang et al.，2016）。本研究中，脑内 *FIH* 表达的上调可能与 HIF-1α 有关，证实了 FIH 与 HIF-1α 快速升高之间可能存在反馈调节机制（Geng et al.，2014）。肿瘤抑制蛋白 p53 是环境压力的通用传感器，可响应各种压力信号，如低氧和紫外线（Koumenis et al.，2001）。在南美白对虾中，低氧可能会增加或减少 *P53* 转录，具体取决于组织类型（Felix-Portillo et al.，2016，Nuñez-Hernandez et al.，2018）。本研究中，洛氏鱥心脏中的 *P53* 基因表达在持续低氧期间先升高后降低，表明该基因在持续低氧期间发挥作用，可能参与细胞周期阻滞、DNA 修复和细胞凋亡。此外，随着低氧程度的增加，由于体内反馈调节的作用，P53 的表达趋向于稳定（Huang et al.，2020）。在持续低氧条件下，心脑中低氧信号分子（如 HIF-1α/HIF-2α/HIF-3α/FIH/VHL/P53）的 mRNA 表达上调，表明低氧信号通路被激活。

第六章 洛氏鱥低氧耐受相关基因克隆及表达分析

第一节 缺氧诱导因子 *HIF* 基因克隆及表达分析

由于富营养化和全球变暖的持续蔓延,缺氧正在成为影响全球水域的环境问题(Vaquer-Sunyer and Duarte,2008)。对于各种生物来说,缺氧会影响细胞相互作用网络和正常发育。一些报告表明,缺氧是塑造动物生理进化的主要力量(Dauer,1993;Rytkönen et al.,2007;Weisberg et al.,2008)。缺氧诱导因子(HIF)在氧稳态和各种组织的先天免疫反应中发挥重要作用(Choi et al.,2013,Semenza and Wang,1992)。它们是由 HIF-α 和 HIF-β 两个亚基(也称为芳基烃受体核转位子,ARNT)组成的异二聚体,属于基本螺旋-环-螺旋(bHLH)-Per-Arnt-Sim(PAS)转录因子家族(Sogawa and Fujii-Kuriyama,1997;Wang and Zhang 1995,Wenger et al.,1997)。在含氧量正常的条件下,脯氨酰羟化酶(PHD)通过对氧依赖性降解结构域(ODD)内的两个保守脯氨酸残基进行羟基化靶向 HIF-α 亚基(Ivan et al.,2001;Fandrey et al.,2006)。这种羟基化的 HIF-α 亚基被 VHL 蛋白识别并进行泛素化,从而确保蛋白酶体介导的降解。在缺氧情况下,HIF-α 亚基会稳定下来并进入细胞核与 HIF-1β 形成异二聚体。这种 HIF 异二聚体继续激活细胞中大量缺氧诱导基因的转录,以应对缺氧环境(Bracken et al.,2003;Kajimura et al.,2005)。只有在缺氧条件下,α 亚基才能与 β 亚基二聚化形成 HIF,从而激活下游氧敏感相关基因的转录,调节生理过程(如血管生成、能量代谢、细胞增殖和凋亡),在缺氧条件下维持生存。

到目前为止,已经在哺乳动物中表征了三种类型的 HIF,即 HIF-1α、HIF-2α 和 HIF-3α(Ratcliffe,2007)。在这三种类型中,HIF-1α 被认为在无氧呼吸中间体的摄氧或输送中起关键作用(Hu et al.,2003;Semenza,2000),并且提出 HIF-2α 在血管生成中和红细胞生成中起重要作用(Hu et al.,2006;Law et al.,2006)。在鱼类中,HIF-1α 已在几个物种中被发现,如青海湖裸鲤(*Gymnocypris przewalskii*)(Cao et al.,2005)和欧洲狼鲈(*Dicentrarchus labrax*)(Terova et al.,2008);HIF-2α 已在底鳉(*Fundulus heteroclitus*)中发现(Powell and Hahn,2002);HIF-3α 已在胭脂鱼(*Myxocyprinus asiaticus*)中

获得（Chen et al.，2012）。此外，HIF-1α 和 HIF-2α 在斑马鱼（Rojas et al.，2007）、草鱼（Law et al.，2006）、虹鳟（Soitamo et al.，2001）、细须石首鱼（*Micropogonias undulates*）（Rahman and Thomas，2007）、团头鲂（Shen et al.，2010）和胭脂鱼（*M. asiaticus*）（Chen et al.，2012）中获得。*HIF-4α* 基因在草鱼中被报道（Law et al.，2006），它被认为与哺乳动物 HIF 高度同源（Law et al.，2006；Powell and Hahn，2002）。如今在一些鲤科鱼类中报道了更多的 *HIF-αs* 基因拷贝（HIF-1αA、HIF-1αB、HIF-2αA 和 HIF-2αB）（Rytkönen et al.，2013）。*HIF-1α* 的表达水平受多种因素影响，如低温（Rissanen et al.，2006）、缺氧（Wang et al.，2021）、天然小分子物质（Liu et al.，2020）和重金属暴露（Sun et al.，2021）。洛氏鱥作为高纬度耐受低氧的鱼类，其 HIF 的相关研究较为缺乏。

一、洛氏鱥 *HIF-1α*、*HIF-2α* 和 *HIF-3α* 基因全长克隆

（一）洛氏鱥 *HIF-1α* 基因全长克隆

克隆得到 *HIF-1α* 基因全长，ORF 编码 773 个氨基酸，序列目前已经提交 NCBI 的 GenBank（MT700411）。HIF-1α 的分子质量和等电点分别为 85.33ku 和 4.98。与其他鱼类结构域类似，洛氏鱥的 HIF-1α 由 6 个结构域组成，包括 HLH、PAS-A、PAS-B、PAS 的 C 端（PAC）、N 端反式激活结构域（N-TAD）和 C 端反式激活结构域（C-TAD）。系统发育分析表明，HIF-1α 的氨基酸序列与团头鲂的氨基酸序列具有最高的同一性（84%）。3D 结构显示，HIF-1α 主要结构域包括一个由 α 螺旋组成的 HLH 结构域、两个 PAS 结构域和一个 HIF-αs 蛋白中的 PAC 结构域（图 6-1、图 6-2）。

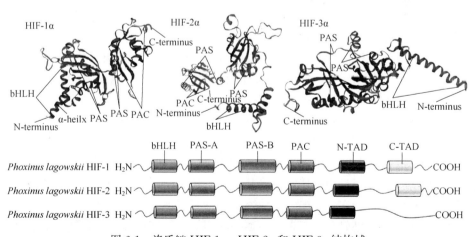

图 6-1　洛氏鱥 HIF-1α、HIF-2α 和 HIF-3α 结构域

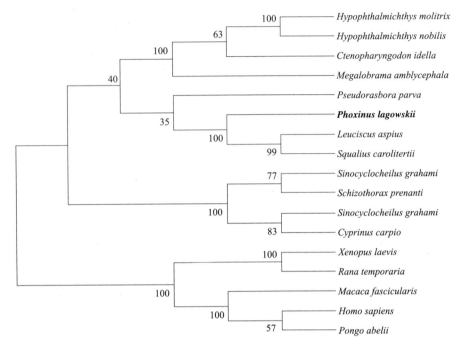

图 6-2 HIF-1α 蛋白系统发育树

(二) 洛氏鲅 *HIF-2α* 基因全长克隆

克隆得到 *HIF-2α* 基因全长，ORF 编码 829 个氨基酸，序列目前已经提交 NCBI 的 GenBank（MT740721）。预测的 HIF-2α 蛋白的分子质量为 92.63ku，等电点为 6.50。与其他鱼类结构域类似，洛氏鲅的 HIF-2α 也由 6 个结构域组成，包括 HLH、PAS-A、PAS-B、PAC、N-TAD 和 C-TAD。系统发育分析表明，HIF-2α 的氨基酸序列与团头鲂的氨基酸序列具有最高的同一性（85%）。蛋白质 3D 结构预测显示，HIF-2α 的主要结构域包括一个 α 螺旋组成的 HLH 结构域、两个 PAS 结构域和一个 HIF-αs 蛋白中的 PAC 结构域（图 6-3）。

(三) 洛氏鲅 *HIF-3α* 基因全长克隆

克隆得到 HIF-3α 的 cDNA 的 ORF 编码 641 个氨基酸，序列目前已经提交 NCBI 的 GenBank（MT740722）。预测的 HIF-3α 蛋白的分子质量为 71.62ku，等电点为 4.26。与其他鱼类结构域类似，洛氏鲅的 HIF-3α 由 5 个结构域组成，包括 HLH、PAS-A、PAS-B、PAC 和 N-TAD，在洛氏鲅 HIF-3α 蛋白中未发现 C-TAD 结构域。系统发育分析表明，HIF-3α 的氨基酸序列与团头鲂的氨基酸序列具有最高的同一性（89%）。3D 结构显示，HIF-3α 主要结构域包括一个由 α 螺旋组成的 HLH 结构域、两个 PAS 结构域和一个 HIF-αs 蛋白中的 PAC 结构域（图 6-4）。

第六章 洛氏鲅低氧耐受相关基因克隆及表达分析

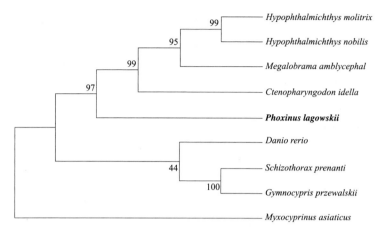

图 6-3　HIF-2α 蛋白系统发育树

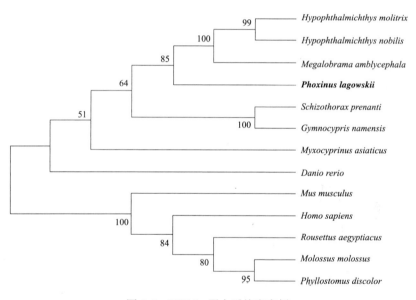

图 6-4　HIF-3α 蛋白系统发育树

洛氏鲅的三种 HIF-α 蛋白含有调节区，包括反式激活结构域、蛋白质稳定化和核易位结构域（Gradin et al.，2002）。它们的功能域具有基本的螺旋-环-螺旋（bHLH）、Per-Sim-ARNT（PAS）-A 和-B，以及核心氧依赖性降解区域（ODD），与其他脊椎动物 HIF-αs 蛋白具有高度序列同一性。一般而言，PAS 结构域与靶基因特异性、反式激活和二聚化有关，在几乎所有研究过的物种中，PAS 结构域中都包含两个 50～100 个碱基组成的序列（PAS-A/B）和疏水区域，并且它们的 PAS 域也具有较高的同源性（Hardy et al.，2012，Kawabe and Yokoyama，2011）。最近的研究发现，在鲤形目的鱼类中，丝氨酸残基的数量增加了鱼对缺氧的抵抗力,其中耐缺氧的鱼在 ODD 结构域中有超过 40 个丝氨酸残基(Chen et al.,2012;Mohindra et al.,2013)。我们发现洛氏鱥 HIF-αs 蛋白含有超过 50 个丝氨酸残基,表明洛氏鱥的耐缺氧潜力。洛氏鱥的 HIF-2α 和 HIF-3α 在 25 位和 24 位有一个半胱氨酸,而 HIF-1α 在 27 位的对应位点是一个丝氨酸。在这项研究中,发现洛氏鱥 HIF-1α、HIF-2α 和 HIF-3α 蛋白在 ODD 结构域具有两个保守的脯氨酸残基,在 C-TAD 结构域具有一个天冬酰胺残基。这与先前的研究一致,即大多数脊椎动物中存在两个保守的脯氨酸残基和一个保守的天冬酰胺残基(Rytkönen et al.,2007)。预测这些残基与蛋白质周转和诱导靶基因转录有关(Lando et al.,2000)。研究还证明了它们在调节 HIF 降解或促进鱼类转录活性方面的功能(Ivan et al.,2001;Law et al.,2006)。LXXLL 序列在洛氏鱥的 HIF-3α 中发现,但在 HIF-1α 和 HIF-2α 亚基中没有发现。其他研究表明,HIF-3α 可能结合 DNA 启动子序列或新的相互作用蛋白质,这些蛋白质与 HIF-1/2α 识别的蛋白质不同(Maynard et al.,2003)。大量研究表明 HIFαs 的 ODDD 和 N-TAD 中含有 LXXLAP 序列在各物种中有所差异。例如,在人类中,HIF-1α 在 ODDD 和 N-TAD 中含有 LXXLAP 中的两个保守的残基(Pro-402 和 Pro-564);在牡蛎和草虾中,ODDD 和 N-TAD 的 LXXLAP 序列中也存在两个脯氨酸残基(Kawabe and Yokoyama,2011;Li and Brouwer,2007)。但在来自杂色鲍(*Haliotis diversicolor*)的 HIF-1α 中,只有一个保守的 Pro407 残基经受羟基化作用,位于 LXXLAP 区域内 ODDD 的 N 端侧(残基 402~407)(Cai et al.,2014)。此外,系统发育树结果表明洛氏鱥的 HIF-1α、HIF-2α 和 HIF-3α 蛋白与鲤科鱼类较为近缘,表明它们之间具有高度进化保守性。洛氏鱥的 HIF-1α 与赤梢雅罗鱼(*Leuciscus aspius*)相似性最高,为 94.95%,与卡氏欧雅鱼(*Squalius carolitertii*)相似性为 91.38%,与鲢相似性为 90.71%。洛氏鱥的 HIF-2α 与草鱼相似性最高,为 91.62%,与鲢相似性为 91.38%,与鳙相似性为 91.14%,与团头鲂相似性为 90.18%。洛氏鱥的 HIF-3α 与团头鲂相似度最高,为 89.75%,与鲢相似性为 89.62%,与鳙相似性为 87.02%。采用 SingalP-5.0 分析表明,洛氏鱥的 HIF-3α 不存在信号肽序列。

二、洛氏鱥 *HIF-1α*、*HIF-2α* 和 *HIF-3α* 基因的组织表达

分析 *HIF-1α*,*HIF-2α* 和 *HIF-3α* 在鳃、心脏、肝脏、脾脏、脑、眼睛、肠和肌肉各组织中的表达情况,发现三个基因在上述组织中均有表达。其中 *HIF-1α* 在眼睛和心脏中表达量较高,但在肌肉和肝脏中表达非常弱($P<0.05$)(图 6-5),在鳃、脾脏、肠道中也有表达。*HIF-2α* 转录物在脾脏、心脏和脑中最为丰富($P<0.05$),在眼睛、鳃和肝脏中也有表达,但是在肌肉

和肠道中的表达程度较低。$HIF-3\alpha$ 在脾脏和肝脏中的表达水平显著高于其他组织，在鳃、眼睛、肌肉中表达量较低（$P<0.05$）。

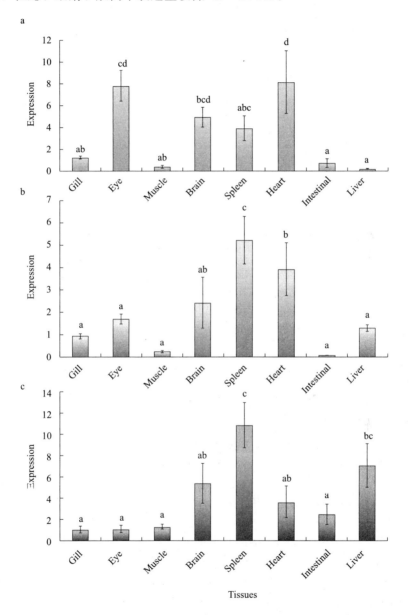

图 6-5 $HIF-\alpha$ 基因通过 qRT-PCR 在常氧下的 8 个组织中测定的相对表达谱

a. $HIF-1\alpha$　b. $HIF-2\alpha$　c. $HIF-3\alpha$

注：gill 代表鳃，eye 代表眼，muscle 代表肌肉，brain 代表脑，spleen 代表脾脏，heart 代表心脏，intetine 代表肠，liver 代表肝脏。图中标有不同字母表示差异显著（$P<0.05$），否则差异不显著。

在正常氧气条件下，$HIF-1α$ 和 $HIF-2α$ 基因在各种组织中的表达已经在一些鱼类中得到证实，在硬骨鱼中表现出组织依赖性差异（Imaoka et al.，2000；Rahman and Thomas，2007）。HIF-α 可以存在于含氧量正常的组织中，以满足细胞能量需求（Stroka et al.，2001）。在翘嘴鳜（*Siniperca chuatsi*）中发现，$HIF-1α$ 在血液、心脏和肝脏中显著高表达，表明翘嘴鳜的 HIF-1α 的主要功能与循环系统密切相关（He et al.，2019）。在胭脂鱼研究中，发现 $HIF-1α$ 和 $HIF-3α$ 在肝脏和性腺中高表达，其次是脾和肌肉（Chen et al.，2012）。在细须石首鱼中，$HIF-1α$ 主要在脑和性腺中表达，其次是心脏、肝脏和肌肉（Rahman and Thomas，2007）。在耐缺氧的淡水鱼草鱼中，在眼睛和肾脏中检测到 $HIF-1α$ 的高表达（Law et al.，2006）。$HIF-1$ 和 $HIF-2$ 在斑马鱼的鳃、卵巢和脑中大量表达（Rojas et al.，2007）。然而，在团头鲂中，$HIF-1α$ 在肝脏中的表达相对高于其他组织，并且 $HIF-2α$ 在大脑和眼睛中表达最多（Shen et al.，2010）。我们之前对卵胎生的鱼类如许氏平鲉（*Sebastes schlegelii*）的研究表明，$HIF-1α$ 和 $HIF-2α$ 呈现出完全不同的组织表达模式（Mu et al.，2015）。本研究在所有受试组织中均检测到了洛氏鱥 HIF-1α、HIF-2α 和 HIF-3α 的 mRNA，这表明所有三个 HIF-α 亚基都在这些组织中发挥作用。三个 $HIF-α$ 亚基在心脏、脾脏和肝脏中的表达明显更高，表明 HIF-α 的主要功能与循环和供能系统密切相关（He et al.，2019）。此外，$HIF-1α$ 在眼睛和心脏中的表达水平最高，而 $HIF-2α$ 和 $HIF-3α$ 在脾脏中的表达水平高于其他组织。上述发现表明，三个 HIF-α 亚基的表达和调控模式具有物种和组织特异性，这可能与这些物种的不同缺氧适应性有关（Rahman and Thomas，2007；Rytkönen et al.，2007）。

三、洛氏鱥 *HIF-1α*、*HIF-2α* 和 *HIF-3α* 基因在不同低氧处理下的表达模式

（一）荧光定量分析 *HIF-1α*、*HIF-2α* 和 *HIF-3α* 基因在不同低氧处理下的表达模式

为了进一步分析 $HIF-αs$ 基因在低氧状态下的表达模式，我们采用 QPCR 的方式分析 $HIF-1α$、$HIF-2α$ 和 $HIF-3α$ 经过低氧胁迫后在肝脏、鳃、脾脏中的表达量（图 6-6）。在肝脏中，$HIF-1α$ 基因表达量在复氧组为最高值，显著高于其他各组（$P<0.05$），但是其他各组与对照组相比均没有显著差异（$P>0.05$）。在鳃中，$HIF-1α$ 的表达量在 0.5h 0.5mg/L 组表达量最高，但是在复氧组表达量最低。在脾脏中 $HIF-1α$ 基因表达量在 0.5h 0.5mg/L 组和 6h 3mg/L 组以及复氧组显著高于对照组和 24h 3mg/L 组（$P<0.05$）。$HIF-2α$ 基因表达量在肝脏的复氧组中最高，但是与对照组相比没有显著差异（$P>$

0.05)。在鳃中，*HIF-2α* 基因表达量在 6h 3mg/L 和 24h 3mg/L 组显著高于其他处理组（$P<0.05$），在 24h 3mg/L 时最高。在脾脏中，*HIF-2α* 基因表达量各组均低于对照组，0.5h 0.5mg/L 组和 6h 3mg/L 组显著低于对照组（$P<0.05$）。肝脏中，*HIF-1α* 基因表达量在 24h 3mg/L 达到了最大值，但是各组与对照组相比均没有显著性差异，且各组间无显著性差异（$P>0.05$）。在鳃中，6h 3mg/L 和 24h 3mg/L 组显著高于对照组和其他各组（$P<0.05$），在 6h 3mg/L 组的值最大。在脾脏中，24h 3mg/L 组为最低值，显著低于复氧组（$P<0.05$）。

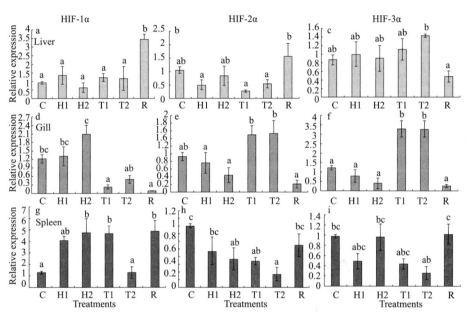

图 6-6 *HIF-α* 基因响应不同低氧条件的表达

a、b 和 c. 肝脏　　d、e 和 f. 鳃　　g、h 和 i. 脾脏

注：平均值±SEM（n=6），不同字母的组之间存在显著差异（$P<0.05$）。C 代表对照组，H1 代表 0.5h 3mg/L，H2 代表 0.5h 0.5mg/L，T1 代表 6h 3mg/L，T2 代表 24h 3mg/L，R 代表复氧组。

肝脏、鳃和脾脏是了解硬骨鱼代谢和呼吸的重要组织（Fatouros et al.，1995；Rui，2014；Tarantino et al.，2013）。尽管许多研究充分证明了 HIF-1α 和 HIF-2α 响应缺氧的功能（Wang et al.，2019），但我们的研究首次开展了低氧处理后洛氏鱥 *HIF-1α*、*HIF-2α* 和 *HIF-3α* 基因的表达模式。肝脏在低氧条件下起到了鱼类体内平衡和能量平衡的关键作用（Li et al.，2018）。在我们的研究中，当洛氏鱥暴露于复氧时，在肝脏中观察到 *HIF-1α* 和 *HIF-2α* 基因表达水平显著增加。这与胭脂鱼的发现一致，即在短期低氧暴露后的复氧组中，*HIF-α* mRNA 表达量显著增加（Chen et al.，2012）。在复氧组中，*HIF-1α*

和 HIF-$2α$ mRNA 水平增加，但 HIF-$3α$ 没有增加，这表明 HIF-$1α$ 和 HIF-$2α$ 替代了 HIF-$3α$ mRNA 对低氧的适应作用。我们的研究结果证实，在较长时间胁迫时，HIF-$3α$ 在肝脏稳态中起重要作用。鳃组织是大多数水生动物最先接触水生环境的器官，因此容易受到环境变化的影响。鱼类的鳃组织对环境波动（如冷应激和低氧）的敏感反应已得到充分证明（Negro and Collins，2017；Zhang et al.，2019）。洛氏鲅中 HIF-$2α$ 和 HIF-$3α$ 基因表达水平在 6h 和 24h 低氧组中始终较高，但在 0.5h 组中不高，表明 HIF-$2α$ 和 HIF-$3α$ 在鳃中容易受到长期低氧的影响。根据 Rahman 和 Thomas（2007）的研究结果，HIF-$2α$ 不参与低氧适应反应的早期阶段。在鳃中，HIF-$1α$ 基因表达在各组处理之间没有显著差异，但在 0.5mg/L 0.5h 暴露时表达显著增加。我们的研究结果补充了一些数据空白，表明在低氧暴露期间 HIF-$1α$、HIF-$2α$ 和 HIF-$3α$ 的不同表达模式（Mu et al.，2015；Shen et al.，2010）。与冷水鱼虹鳟相比，虹鳟在 3mg/L 6h 缺氧条件下 HIF-$1α$ 显著增加（Hou et al.，2020），但在洛氏鲅中没有显著表达；这种模式可能反映了低氧反应的物种特异性差异。本研究证实，短期缺氧的鳃中 HIF-$1α$ 的表达水平较高，而长期缺氧组中 HIF-$2α$ 的表达水平较高。根据前人研究，HIF-$1α$ mRNA 表达水平比 HIF-$2α$ 更早升高。因此，假设 HIF-$1α$ 作为急性缺氧敏感调节剂，而 HIF-$2α$ 作为长期缺氧反应的介质（Rahman and Thomas，2007）。硬骨鱼脾脏是一个独立的器官，含有参与新红细胞合成的红细胞生成组织，从而维持鱼类的造血功能（Wells and Weber，1990；Zhao et al.，2015）。本研究分析了来自洛氏鲅脾脏的三个 HIF-$α$ 亚基在各种低氧处理中的表达，结果表明 0.5mg/L 0.5h、3mg/L 6h 和复氧组 HIF-$1α$ 在脾脏中的表达显著增加。短期低氧时 HIF-$1α$ 表达增加可能与脾脏对短期低氧敏感有关，且红细胞生成过程可能涉及低氧。需要进一步开展 HIF-$2α$ 和 HIF-$3α$ 的作用研究，以证实这两个基因在洛氏鲅脾脏中的表达下调可能表明它们对调节低氧耐受性的重要性较差。我们的结果与其他鱼类一致，包括团头鲂（Shen et al.，2010）和鲈（Terova et al.，2008），表明在短期和长期缺氧暴露期间，只有一个亚基（HIF-$1α$ 或 HIF-$2α$）的转录水平受到调节并显著上升。HIF-$1α$ 和 HIF-$2α$ 在鳃和脾脏中的差异表达模式证实了这两个亚基可能具有独特和互补的作用（Mohindra et al.，2013）。在星丽鱼（*Astronotus ocellatus*）肝脏中 HIF-$1α$ 的 mRNA 转录水平在恢复过程中降低，这表明该基因的稳定是鱼类维持体内平衡的一种策略（Baptista et al.，2016）。

(二) 荧光原位杂交分析 HIF-$1α$、HIF-$2α$ 和 HIF-$3α$ 基因在不同低氧处理下的表达模式

荧光原位杂交结果表明 HIF-$αs$ 的表达主要在洛氏鲅的肝细胞质中。结合

qRT-PCR 分析的结果，这些结果表明 HIF-3α 在肝脏中的表达高于 HIF-1α（彩图 11）和 HIF-2α（彩图 12）。因此，HIF-3α 可能在急性缺氧反应中起主要作用。在未进行低氧处理的对照组中，HIF-3α 的荧光信号强度较弱（彩图 13）。然而，在 3mg/mL 低氧处理 24h 后，肝组织中 HIF-3α 的荧光信号强度显著增加。

该研究首次发现洛氏鲅在肝脏缺氧 24h 期间上调 HIF-3α（但不是 HIF-1α 和 HIF-2α）。qRT-PCR 分析和 FISH 的结果显示，HIF-3α 在肝脏中的表达高于 HIF-1α 和 HIF-2α，证实 HIF-3α 在肝脏稳态中起更重要的作用。FISH 的主要优点之一是它允许在组织或细胞水平上定位特定基因，这种组织或细胞特异性反应为理解分子和组织学过程之间的关系提供了线索（Park et al.，2008）。FISH 结果显示 HIF-3α 在正常氧气条件下主要存在于肝细胞的细胞质中。缺氧 24h 后，HIF-3α 的 RNA 表达明显高于对照组，这与我们的 qRT-PCR 结果一致。此外，在大多数细胞的核区也观察到 HIF-3α 的表达。之前对斑马鱼的研究表明，斑马鱼 HIF-3α 定位在细胞核中，其中 EGFP 被标记并转染到培养细胞中，这与我们的结果有些相似（Zhang et al.，2012）。我们推测这可能是由于 HIF-3α 在细胞质和细胞核之间的动态分布导致。HIF-1α 在细胞质中积累并转移到细胞核（Zhang et al.，2012）。因此，需要进一步研究以确定 HIF-3α 亚细胞定位，从而更深入地了解 HIF-3α 在硬骨鱼中的功能。

（三）HIF 过表达后 CIK 细胞分析

HIF-3α 在 CIK 细胞中过表达后，细胞中 P53、Glut1、VEGF、VHL 和 FIH 的表达水平如图 6-7 所示。HIF-3α 的过表达显著增加了 P53、Glut1、VEGF 和 VHL 的表达水平（分别是对照组的 1.43、1.36、1.32 和 1.47 倍）（$P<0.05$），而 FIH 的表达水平没有显著变化。

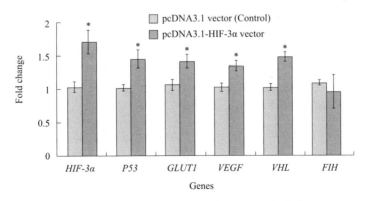

图 6-7 RT-qPCR 分析 HIF-3α 过表达后 CIK 细胞中 P53、Glut1、VEGF、VHL 和 FIH 的表达水平

注：*代表 $P<0.05$。

HIF 是介导细胞适应低氧的转录因子（Rankin and Giaccia，2008）。许多 HIF 调节基因参与葡萄糖转运、糖酵解、红细胞生成、血管生成、血管舒张和呼吸速率，并共同发挥作用以最大限度地减少氧气在细胞、组织和全身水平上的影响（Chen et al.，2012）。例如，鲱形白鲑（*Coregonus clupeaformis*）中第一次观察到的 *HIF-1α* mRNA 响应低氧的变化是在受精后第 38 天下调，这种反应持续到受精后第 83 天，该研究认为在特定时期，这种 *HIF-1α* mRNA 的减少与 HIF-1 靶基因 mRNA 水平的增加是具有相关性的，而这些靶基因包括 *VEGF*（血管内皮生长因子，vascular endothelial growth factor）和 *EPO*（促红细胞生成素，erythropoietin）（Whitehouse et al.，2019）等。尽管 HIF-1α 和 HIF-2α 作为低氧转录反应的主要调节因子的作用已经清楚，但 HIF-3α 在低氧条件下的作用及其作用方式还不清楚（Zhang et al.，2012）。本研究的低氧数据表明，与 HIF-1α 和 HIF-2α 相比，HIF-3α 对洛氏鱥低氧应激状态的变化更为明显。有报道表明，低氧会激活小鼠和大鼠中 *HIF-3α* mRNA 的表达（Pasanen et al.，2010），因此，我们推测 HIF-3α 在洛氏鱥的低氧环境中发挥更大的作用。为了验证这一假设并更好地了解 HIF-3α 在鱼类缺氧反应中的作用，我们构建了 pcDNA3.1-HIF-3α 表达质粒并将其转染到 CIK 细胞中。在 *HIF-3α* 过表达后检测到已知与缺氧相关的基因的表达水平，包括 *P53*、*Glut1*、*VEGF*、*VHL* 和 *FIH* 等。我们的研究结果表明，*P53*、*Glut1*、*VHL* 和 *VEGF* 的表达水平被显著上调，而 *FIH* 的表达水平没有显著变化。许多研究表明，HIF-α 蛋白在低氧癌细胞中稳定促进了肿瘤进展，而且该过程是通过诱导编码 VEGF 和 Glut 1 的特定靶基因的局部表达的（Zhang et al.，2012）。有报道分析了 *HIF-3α* siRNA 对已知低氧诱导因子靶基因 *Glut1* 表达的影响，并显示出 18%～54% 的显著下调（Heikkilä et al.，2011）。我们的研究证明 HIF-3α 调节了洛氏鱥的 Glut1。VHL 病是一种遗传性癌症综合征，其特征是发展为过度产生缺氧诱导型 mRNA，如 VEGF 的高血管肿瘤（Ivan et al.，2001）。此外，也有研究证实在缺氧 3h 后，肝脏中 VEGF 的表达水平显著增加，这反映了我们的结果（Baptista et al.，2016）。我们的研究结果表明，*HIF-3α* 的表达调节缺氧相关的靶基因，证实 *HIF-3α* 基因的转录水平可能是低氧环境中有用的分子生物标志物。因此，我们认为洛氏鱥还通过在低氧耐受机制中上调 HIF-3α 的表达来应对缺氧损伤，从而调节糖脂代谢和血管生成。

第二节 *P53* 基因克隆及表达分析

鱼类在长期演化历程中受到不同溶氧浓度水体的自然选择，各鱼类物种对溶解氧适应力也存在显著差异,为研究生物体对低氧环境的适应和调控机制提供了绝佳的材料。当鱼类处于缺氧状态时,氧化和抗氧化系统之间的平衡被打破,导致活性氧(ROS)产生增加和随后的氧化应激(Comte and Olden,2017)。这些可能是由于线粒体电子传递链的细胞色素减少并泄漏到残余氧分子中而导致的 ROS 水平增加(Lushchak and Bagnyukova,2006)。过量的 ROS 会导致蛋白质结构改变或生物活性丧失、DNA 链断裂、DNA 位点突变、DNA 双链畸变以及原癌基因和肿瘤抑制基因突变,最终导致氧化损伤和代谢障碍(Ziech et al.,2008)。为了抵消 ROS 的有害影响,大多数生物体已经开发出可以预防、改善或修复 ROS 造成的损害的防御机制(Metcalfe et al.,2010)。暴露于低氧环境和升高的 ROS 水平可以协同作用导致 DNA 损伤,导致细胞凋亡。与此同时,硬骨鱼可以通过分子调控机制有效地利用低水平溶解氧,部分鱼类已经进化出相应的应对策略,让它们在低氧的条件下生存(Wang et al.,2017)。而肿瘤抑制基因 P53 在细胞对氧化应激的反应中起关键而复杂的作用。

肿瘤抑制基因 P53 是人类和其他哺乳动物细胞周期进程中重要的负调控因子,且 P53 蛋白在转录激活、DNA 合成和修复以及程序性细胞死亡中起着重要的作用(Caelles et al.,1994)。在许多水生生物的研究中发现,肿瘤抑制蛋白 P53 是环境压力的通用传感器,并且作为转录因子,在暴露于引起 DNA 损伤的试剂的细胞的存活或死亡中起着至关重要的作用,同时参与 DNA 修复、细胞周期阻滞、衰老和凋亡等各种生命活动(Qi et al.,2013;Huang et al.,2020)。为了应对低水平的氧化应激,P53 主要发挥抗氧化作用。P53 还可以通过调节细胞代谢降低细胞内 ROS 的水平。在这种情况下,TP53 诱导的糖酵解和凋亡的调控因子(TP53-induced glycolysis and apoptosis regulator gene,TIGAR)被 P53 诱导表达,从而减缓糖酵解并促进 NAPDH 的产生以降低 ROS 水平(Bensaad et al.,2006)。此外,磷酸甘油酸变位酶(Phosphoglucomutase,PGM)可被 P53 抑制,导致线粒体氧化呼吸所需的丙酮酸减少,从而减少 ROS 的产生(Bensaad et al.,2007;Matheu et al.,2007)。P53 蛋白也作为上游信号的整合子发挥作用,然后作为信号转导网络中的中心节点,该响应最小化可导致癌症或其他病症的突变(Levine et al.,2004)。

作为一种转录因子,P53 对各种压力信号作出反应,如低氧、紫外线辐射、活性氧和突变体(Koumenis et al.,2001)。此外,P53 蛋白的调节一直是癌症领域关注的重点,它对于进一步了解人类、小鼠、兔子和其他哺乳动物的癌症相关调控至关重要(Matlashewski et al.,1984;Meira et al.,2008;Goas,1997;Levine et al.,2004)。近来,P53 在水生动物中的功能分析是一个热门研究内容,特别是在一些水产养殖物种中,如鲤、罗非鱼(Mai et al.,2010)和斑节对虾

(Dai et al., 2016)。此外，*P53* 基因已在多种鱼类中被鉴定，如虹鳟（Fromentel et al., 1992）、青鳉（*Oryzias latipes*）（Krause et al., 1997）、斑马鱼（Cheng et al., 1997）、川鲽、大西洋鲑和草鱼（Huang et al., 2017）。

在这些研究中已经确定 P53 蛋白包含三个主要功能结构域：反式激活结构域、DNA 结合结构域和涉及寡聚化的结构域。目前，关于鱼类 *P53* 基因应对低氧胁迫后的组织表达分析和不同氧气胁迫后的表达量研究严重不足，对于冷水性鱼类的研究更加缺乏，急需要补充此方向的信息来为鱼类低氧耐受性提供更多的基础数据。

一、洛氏鱥 *P53* 基因全长克隆

（一）洛氏鱥 *P53* 基因全长克隆

利用 SMART-RACE 技术获得洛氏鱥 *P53* 的全长。洛氏鱥 *P53* cDNA 序列为 1878bp，包括 219bp 5′非翻译区，1116bp 的 ORF 和 543bp 的 3′UTR。3′UTR 区域包含典型的尾信号 AATAAA 和一个 poly-A 尾。ORF 编码 371 个氨基酸，分子质量为 41.22ku，理论等电点为 7.38。洛氏鱥 P53 核苷酸和推断的氨基酸序列已提交 NCBI（MZ821024）。分析表明 P53 蛋白在 N 端具有疏水性，在 C 端具有亲水性。与其他物种 P53 蛋白的序列比较表明，洛氏鱥 P53 蛋白共有 5 个高度保守的结构域。在 SWISS-MODEL 上预测了洛氏鱥 P53 蛋白的三维结构（图 6-8）。结果表明，洛氏鱥 P53 蛋白具有保守的三级结构，具有典型的 α-螺旋和 β-折叠。P53 蛋白是一种防止肿瘤转化的四聚体转录因子。寡聚化似乎对 P53 的肿瘤抑制活性至关重要。洛氏鱥的 P53 包含几个结构域，即 N 端反式激活结构域（TAD）、富含脯氨酸的结构域（PD）、DNA 结合结构域（DBD）。其中 TAD（残基 4~14）对 P53 调控至关重要，OD 结构域使得 P53 形成四聚体（残基 289~331）。四聚化结构域对于 DNA 结合、蛋白质-蛋白质相互作用、翻译后修饰和 P53 降解至关重要。采用 SingalP-5.0 分析表明，洛氏鱥的 P53 不存在信号肽序列。采用 NetPhos 3.0 预测洛氏鱥 P53 的翻译后修饰情况，发现洛氏鱥 P53 的编码蛋白有 44 个磷酸化位点，其中有 29 个丝氨酸位点、15 个苏氨酸位点。带负电荷的残基总数（Asp+Glu）是 52，带正电荷的残基总数（Arg+Lys）是 52，没有跨膜结构域。

（二）洛氏鱥 P53 的系统进化分析

为了探究不同物种间 P53 在进化上的关系，我们构建了系统进化树。系统进化树显示，这些蛋白质可分为三大类，即硬骨鱼类、两栖类和哺乳类（图 6-9）。结果表明，洛氏鱥 P53 与黑头呆鱼（*Pimephales promelas*）的亲缘关系最近，为 93.89%；其次是草鱼，相似度为 89.69%，和团头鲂的亲缘关系也较高，为 88.93%。

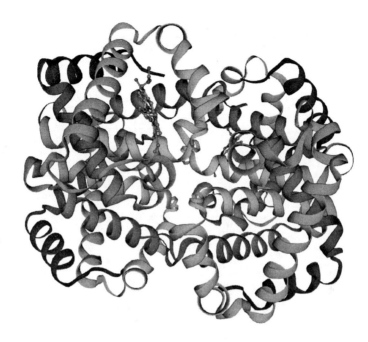
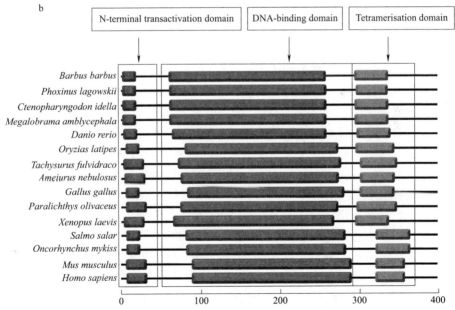

图 6-8 洛氏鱥 P53 蛋白
a. 三维结构 b. 与其他物种的结构域的比较
注：不同的颜色代表不同的结构域。

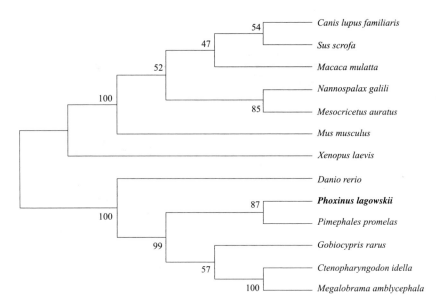

图 6-9 基于洛氏鱥 P53 氨基酸序列的 NJ 树

（三）综合分析

P53 转录因子家族（P53、P63 和 P73）（Irwin and Kaelin，2001；Kaghad，1997）被认为是脊椎动物中重要的基因家族之一。其功能的轻微改变通常会强烈影响细胞命运（Amelio and Melino，2015）。P53 家族成员的整体结构域是非常保守的，其编码几种不同亚型的 mRNA（Ampofo et al.，2010）。P53 在鱼类免疫（Guo et al.，2017）、卵巢发育（He et al.，2021）以及对环境中的化学物质的反应（Li et al.，2008）中发挥着至关重要的作用。此外，P53 是环境压力的通用传感器，作为转录因子，它在调节基因表达中发挥关键作用，介导响应不同条件的生长停滞和细胞死亡。低氧也会诱导 P53 的积累（Chen，2003；Nakano et al.，2014）。在这项研究中，我们报告了洛氏鱥 p53 的完整 cDNA 序列和特征。洛氏鱥 P53 序列与其他硬骨鱼具有高度同源性（47%～88.68%）。此外，洛氏鱥 P53 蛋白序列具有主要特征，如疏水性的 N 末端，亲水性 C 末端和倒数第二个丝氨酸残基。典型的 DNA 结合和四聚结构域证实了其作为转录因子的调节功能（Qi et al.，2013）。此前，研究表明 P53 在进化中高度保守，即它包含两个簇，酸性疏水氨基酸、带电荷的亲水 C 末端和磷酸化丝氨酸残基（Rui et al.，2010）。在 C 末端，保守的四聚结构域、核输出信号以及保守的核定位信号序列的存在表明洛氏鱥 P53 同源物通过其寡聚化和亚细胞定位进行调节（Farcy et al.，2008）。洛氏鱥 P53 的 TAD 和 TD 与草鱼和团头鲂具有较高的同一性，但与其他物种（青鳉、小鼠和人）差异较大，表明 P53 的突变主要发生在这个区域（Levine，1997；May，1996）。同时，洛氏鲅 P53 与黑头呆鱼、草鱼、团头鲂的相似度极高，表明 P53 具有较高的保守性。

二、洛氏鲅 *P53* 基因组织表达结果

在分析鳃、脑、眼睛、心脏、肝脏、肌肉、肠和脾等组织中的洛氏鲅 P53 表达（图 6-10）后发现，洛氏鲅 *P53* 基因在所有检测的组织中均有表达。这与大部分水生动物的研究结果是一致的，表明 P53 在各组织中的广泛作用。但在这些组织中表达水平存在显著差异，洛氏鲅 *P53* 基因表达在心脏和脾脏中最高（$P<0.05$），在脑中最低（$P<0.05$），且 P53 在鳃、脑、眼、肌肉、肠中也有表达。与大菱鲆研究结果不同，大菱鲆 P53 在肝脏中表达量最高，这可能与肝脏是负责吸收和储存的主要器官，肝脏对环境压力敏感有关（Huang et al.，2020）。在对虾的研究中发现，P53 在肌肉、鳃、血细胞、心脏和肠中均有表达，而在肝胰腺中 mRNA 的相对表达水平显著更高。肝胰腺是负责吸收和摄入物质的主要器官，猜测水产动物肝胰腺可能对水生生物中的缺氧反应敏感（Sun et al.，2016）。鳜中 *P53* 基因在鳃和肾脏中表达量最丰富，在肌肉和脑中表达量较低，认为脾脏是 ISKNV（鳜传染性脾肾坏死病毒）复制的靶器官，可用于评估免疫相关基因的表达（Guo et al.，2016）。此外，在斜带石斑鱼（*Epinephelus coioides*）的肠道和肝脏中，*P53* mRNA 表达水平显著高于肌肉、鳃、血细胞、心脏、肾脏和脾脏等组织（Qi et al.，2013）。在稻田鳗（*Monopterus albus*）的研究中，*p53* mRNA 表达水平在卵巢中最高，其次是脾脏（Guo et al.，2021）。结合上述数据，我们得出结论，p53 在组织中的表达模式因物种而异。

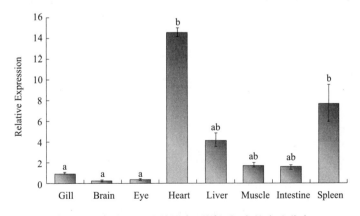

图 6-10 洛氏鲅 *P53* 基因在不同组织中的表达分布

注：gill 代表鳃，brain 代表脑，eye 代表眼，heart 代表心脏，liver 代表肝脏，muscle 代表肌肉，intetine 代表肠，spleen 代表脾脏。图中标有不同字母表示差异显著（$P<0.05$），否则差异不显著。

（一）洛氏鱥 P53 基因经低氧胁迫后脑、心脏、肌肉、肝脏和脾脏中的表达分析

如图 6-11 所示，洛氏鱥 P53 经短期低氧后，脑中的表达量以对照组最高，处理组显著低于对照组（$P<0.05$）。在心脏中，对照组显著高于其他低氧处理组（$P<0.05$），在 24h 3mg/L 胁迫组表达量最低。在肌肉和肝脏中，6h 3mg/L 胁迫组表达量最高，且显著高于其他低氧处理组（$P<0.05$）。在脾脏中，P53 基因表达量在 24h 3mg/L 胁迫组最高，显著高于对照组和其他处理组（$P<0.05$）。

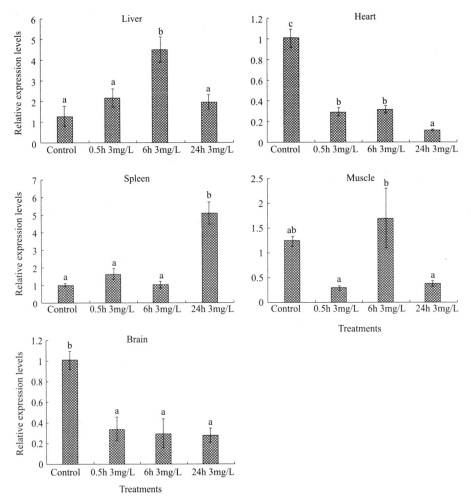

图 6-11　洛氏鱥 P53 在短期低氧胁迫下不同组织中表达水平的变化

注：liver 代表肝脏，heart 代表心脏，spleen 代表脾脏，muscle 代表肌肉，brain 代表脑。图中标有不同字母表示差异显著（$P<0.05$），否则差异不显著。

第六章 洛氏鲅低氧耐受相关基因克隆及表达分析

如图 6-12 和图 6-13 所示，经长期低氧胁迫后，洛氏鲅脑组织中，*P53* 基因在持续低氧组随着胁迫时间延长而上升，在 10d 时达到最大值，在昼夜低氧胁迫下，处理 4d 后，表达量显著降低（$P<0.05$），低于对照组和处理 2d。洛氏鲅心脏中，*P53* 基因在持续低氧和昼夜低氧处理下表达量下降，各组的表达量均显著低于对照组（$P<0.05$）。*P53* 基因在持续低氧处理时，肌肉中的表达量在 6d 达到最高值，显著高于其他低氧处理组（$P<0.05$）。在昼夜低氧处理时，*P53* 基因呈现先上升后下降的趋势，在处理 6d 达到最高，随后下降。肝脏中，*P53* 基因的表达量在持续低氧处理时，4d 时达到最大值，而在昼夜低氧处理时 2d 达到最大值。在脾脏中，*P53* 基因的表达量在持续低氧处理时，各组的表达量均显著高于对照组（$P<0.05$）；而昼夜低氧处理中，脾脏 *P53* 基因表达量在 6d 时最高，显著高于对照组（$P<0.05$）。

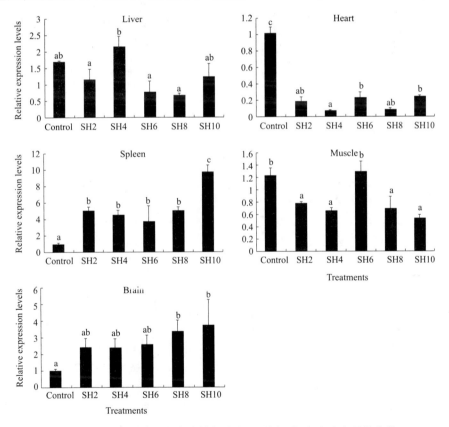

图 6-12　洛氏鲅 P53 在持续低氧胁迫下不同组织中表达水平的变化

注：Control 为对照组，SH2 为持续低氧 2d 组，SH4 为持续低氧 4d 组，SH6 为持续低氧 6d 组，SH8 为持续低氧 8d 组，SH10 为持续低氧 10d 组。图中标有不同字母表示差异显著（$P<0.05$），否则差异不显著。

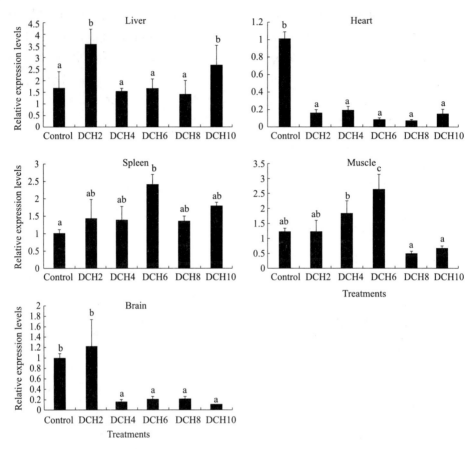

图 6-13 洛氏鲅 P53 在昼夜低氧胁迫下不同组织中表达水平的变化

注：Control 为对照组，DCH2 为昼夜低氧 2d 组，DCH4 为昼夜低氧 4d 组，DCH6 为昼夜低氧 6d 组，DCH8 为昼夜低氧 8d 组，DCH10 为昼夜低氧 10d 组。图中标有不同字母表示差异显著（$P<0.05$），否则差异不显著。

P53 蛋白被多种压力激活，包括 DNA 损伤、缺氧、氧化还原电位的变化、三磷酸核糖核苷池的减少、黏附和几种癌基因的表达（Haupt，1999）。氧化应激已被广泛报道是各种细胞凋亡的重要诱导因素（Franco et al.，2009）。P53 被广泛地翻译后修饰，包括磷酸化和乙酰化，进一步调节其稳定性和活性。在氯化钴（$CoCl_2$）诱导的缺氧中，P53 在神经元细胞中被激活（Yoo et al.，2019）。持续低氧处理后脑中 P53 的表达明显高于对照组，可能是大脑依赖线粒体来产生能量，这些细胞需要密切协调它们的新陈代谢以维持基本的身体功能。持续的缺氧会导致大脑中的氧化应激，破坏线粒体并增加 P53 的表达（Stefanatos and Sanz，2018）。因此，面对不同的缺氧条件，洛氏鲅大脑中的 P53 受到不同的调节。

第六章 洛氏鳄低氧耐受相关基因克隆及表达分析

与对照组相比，在短期和长期缺氧条件下洛氏鳄心脏中 P53 的表达降低。心脏中 *P53* 表达降低可通过减少 P53 的积累来改善心室重构，从而防止心肌细胞凋亡（Naito et al.，2010）。此外，有研究报道缺氧导致骨骼肌缺血损伤，并且缺氧还通过许多转录因子的变化抑制肌肉生成，在此过程中，*P53* 表达增强以在补偿这种损伤中发挥重要作用（Beyfuss and Hood，2018）。活性氧诱导的细胞因子释放可能有助于肝损伤的发病，并且它也被证明可以诱导 P53 活化（Yahagi et al.，2004）。洛氏鳄肝脏中的 P53 在 6h 低氧处理、4d 持续低氧处理和 2d 循环低氧处理时高表达，证实 P53 可以在缺氧期间在肝脏中被激活。大量的研究表明，脾脏维持鱼类的生理稳态和造血功能（Zhao et al.，2017）。低氧处理后洛氏鳄脾脏中 P53 的表达显著增加。*P53* 在缺氧条件下刺激造血细胞的成熟（Ganguli，2002），脾组织中 *P53* 的表达升高可能与造血功能有关，以提高鱼类的携氧能力（Wells et al.，1989）。

（二）FISH 检测缺氧肝细胞中洛氏鳄 *P53* 基因的表达

为了表征细胞位置，进行了洛氏鳄 P53 的 FISH 分析。如彩图 14 所示，洛氏鳄 P53 信号分布在肝细胞的细胞质中。与 qRT-PCR 结果一致，与对照组相比，肝细胞中洛氏鳄 P53 的信号在 6h 缺氧组和 2d 时昼夜低氧组增强。此外，在 2d 时，持续低氧组的肝细胞中也检测到了相对强的信号。

肝细胞可以作为鱼体内氧化应激的良好指标（Mustafa et al.，2015）。我们在 24h 缺氧、2d 持续低氧处理和循环缺氧后从洛氏鳄中选择肝细胞用于 FISH 分析。FISH 结果显示洛氏鳄 P53 主要在肝细胞的细胞质中表达。这是通过原位杂交首次证明 P53 主要存在于洛氏鳄肝细胞的细胞质中。FISH 结果的荧光强度与洛氏鳄 P53 的表达模式一致。

第三节 *HO* 基因的片段克隆及表达分析

由于自然条件下溶解氧的波动，水生缺氧被认为是影响养殖鱼类生存的关键因素之一。有大量研究考察了参与适应缺氧条件过程的相关基因，鱼类对缺氧条件的适应程度因物种而异（Cai et al.，2014；Geng et al.，2014；Okamura et al.，2018）。在高纬度地区，冬季冰雪覆盖水域持续低氧是一种普遍现象。在高纬度地区，水中浮游植物白天进行光合作用，在夏季经常发生循环缺氧（Williams et al.，2019；Wang et al.，2021）。缺氧会导致蛋白质和细胞因子发生变化，进而影响鱼类的行为、生理、生化和分子变化，最终改变整个种群的生物学特征（Farrell and Richards，2009；Xiao，2015）。血红素加氧酶（HO）是一种催化血红素转化为胆红素并同时产生铁和一氧化碳（CO）的酶，是一种通用、高活性和敏感的酶，在植物中受到了广泛研究（Shekhawat et al.，2010）。血红素加氧酶分为三种类型：氧应激诱导型（HO-1）型、组成型（HO-2）和尚未明确的 HO-3 型，这些酶在铁代谢和维持血红素的稳定状态等生物过程中是必不可少的，是血红素分解代谢的限速酶，参与代谢可形成胆绿素、一氧化碳和亚铁离子。HO-1 是由缺氧诱导的，而 HO-2 是在成分中表达的，但有证据表明 HO-2 参与了缺氧条件下鱼类生物代谢的调节（Nilsson and Renshaw，2004；Rimoldi et al.，2012）。在动物中，HO-1 于 1968 年首次在大鼠肝脏中发现，是血红素加氧酶 HO 的诱导型（Tenhunen et al.，1968；Bauer et al.，2008）。而哺乳动物 HO-2 是 Ca^{2+} 敏感钾（BK）通道的一部分（Williams et al.，2004，Munoz-Sanchez and Chanez-Cardenas，2014）。

大量研究证实，虽然 HO-1 和 HO-2 的结构和机制基本相同，但这两种酶的调节和表达差异较大（Ayer et al.，2016；Munoz-Sanchez and Chanez-Cardenas，2014）。HO-1 是由过量的血红素以及一些非血红素压力源诱导的，如氧化应激、感染和暴露于各种异源物质等。HO-2 在所有组织和细胞类型中组成型表达，在大脑和睾丸中含量最高（Maines et al.，1997；Maines et al.，2005）。在正常氧气条件下，Ca^{2+} 敏感钾通道被一定浓度的一氧化碳所打开，一氧化碳由 HO 引起的血红素解离产生（Riesco-Fagundo et al.，2001，Yi and Ragsdale，2007，Yi et al.，2009）。然而，在缺氧条件下，血红素与 Ca^{2+} 敏感钾通道结合并保持关闭。近年来，逐渐发现 HO-2 在氧传感器中的作用和应对氧气变化的细胞保护功能。

为了更深入地了解高纬度鱼类血红素加氧酶的潜在功能，笔者筛选鉴定出洛氏鱥 HO 基因序列，检测了 HO 基因在各组织中的表达分布情况，并对 HO 基因在不同氧气浓度下脾、肝、脑、鳃和心脏组织中的表达变化进行分析，进一步探讨了 HO-1 和 HO-2 基因在低氧耐受过程中的调控作用。

一、洛氏鱥 *HO-1* 和 *HO-2* 基因基本信息及同源性比较

如图 6-14 所示，获得的洛氏鱥 HO-1 基因长度为 1177bp，ORF 编码 268 个氨基酸。预测的 HO-1 蛋白的分子质量为 30.34ku，理论等电点为 6.33。与其他鱼类结构域类似，洛氏鱥的 HO-1 包含一个 N 端结构域、一个 Heme oxygenase 结构域（残基 11～215），一个 C 端结构域。洛氏鱥 HO-1 不含有跨膜结构域，三维结构分析表明 HO-1 由 α-螺旋和不规则螺旋结构组成。带负电荷的残基总数（Asp+Glu）是 34，带正电荷的残基总数（Arg+Lys）是 32。洛氏鱥 HO-1 的编码蛋白有 28 个磷酸化位点，其中有 18 个丝氨酸位点，10 个苏氨酸位点。洛氏鱥的 HO-1 与鲤鱼 HO-1 相似度为 91.42%，与犀角金线鲃（*Sinocycloheilus rhinocerous*）HO-1 相似度为 90.67%。通过进化分析发现 HO-1 和 HO-2 分为两支（图 6-15），其中 HO-1 与鲤鱼、鳙浪白鱼（*Anabarilius grahami*）、团头鲂和草鱼等聚为一支。HO-2 与野鲮（*Labeo rohita*）、安水金线鲃（*Sinocycloheilus anshuiensis*）和团头鲂等聚为一支。

第六章 洛氏鱥低氧耐受相关基因克隆及表达分析

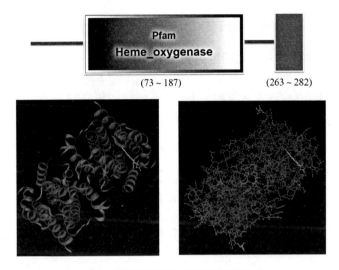

图 6-14 洛氏鱥 HO-1 基因的结构域分析

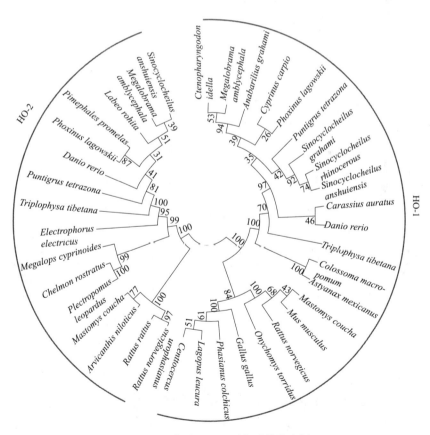

图 6-15 洛氏鱥 HO 蛋白系统发育树

二、洛氏鱥 *HO-1* 和 *HO-2* 基因的组织表达

（一）洛氏鱥 *HO-1* 基因的组织表达

洛氏鱥 *HO-1* 基因在所有检测的组织，包括心脏、肝脏、脾脏、肌肉、肠道、鳃、眼和脑中都有表达（图6-16）。在脾脏中表达量最高，在肌肉中表达量其次。在肠道、心脏、鳃、眼和脑中均有表达。在肝脏中表达量最低。

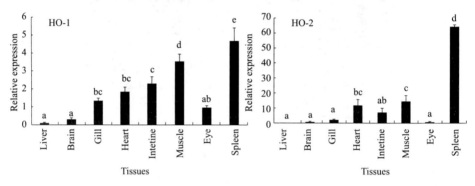

图6-16 洛氏鱥 *HO* 基因的组织表达结果

注：liver代表肝脏，brain代表脑，gill代表鳃，heart代表心脏，intetine代表肠，muscle代表肌肉，eye代表眼，spleen代表脾脏。图中标有不同字母表示差异显著（$P<0.05$），否则差异不显著。

（二）洛氏鱥 *HO-2* 基因的组织表达

洛氏鱥 *HO-2* 基因在所有检测的组织，即心脏、肝脏、脾脏、肌肉、肠道、鳃、眼和脑中均有表达。与 *HO-1* 基因类似，*HO-2* 基因也在脾脏中表达量最高。其次是肌肉、心脏、肠道、鳃、眼、脑。在肝脏中表达量最低。

（三）综合分析

HO-1 和 *HO-2* 基因在所有检测的组织中均有表达。其中 *HO-1a* 的转录本在脾脏、肌肉、肠、心脏和鳃中强烈表达；除此之外，在大多数其他组织中仅检测到较低水平。我们发现 HO-2a 在洛氏鱥脾脏、肌肉和心脏中含量最高，并且在脑、鳃和肝脏中也有表达。在金鲳（*Trachinotus ovatus*）的组织分布分析表明，*HO-1* 在心脏、肝脏和脾脏中的表达水平相对较高（Xie et al.，2020）。军曹鱼（*Rachycentron canadus*）的一项研究表明，*HO* 在所有测试的组织中都有表达，在鳃中表达最高（Xie et al.，2021）。在团头鲂中，*HO* 在脑、心脏和肝脏组织中的表达较高，而在鳃和脾中的表达较低（Guan et al.，2017）。此外，我们的基因表达模式结果与其他鱼类的组织分布分析结果基本一致，表明心脏和脾脏是 *HO-2a* 表达的主要部位，通常表达量超过大脑（Xie et al.，2020，Lu et al.，2021，Xie et al.，2021）。以上结果均表明，*HO* 基因在各组织中的表达具有差异性，且表现为物种特异性。

三、洛氏鲅 *HO-1* 和 *HO-2* 基因在不同低氧处理下的表达模式

（一）*HO-1* 和 *HO-2* 基因在短期胁迫后的表达模式

我们测定了 *HO-1* 和 *HO-2* 基因在肝脏、脾脏、脑、鳃和心脏中的低氧后的相对表达模式（图6-17）。结果发现，HO-1 在肝脏中的3mg/L下的6h和24h均显著上调（$P<0.05$），且在复氧组中表达量也显著高于其他低氧处理组（$P<0.05$），在6h的3mg/L组中达到最高值。在肝脏中，HO-2 在3mg/L处理后的6h和24h，以及复氧组中均显著高于其他各组（$P<0.05$），在6h的3mg/L组中达到最高值。在脾脏中，HO-1 在低氧处理后显著低于对照组（$P<0.05$）。低氧处理后，脾脏 HO-1 在3mg/L的6h组内达到最高值，显著高于其他低氧处理组（$P<0.05$），在0.5h 3mg/L组中达到最低值。此外，脾脏HO-2 在0.5mg/L的0.5h组中达到最高值，显著高于其他低氧处理组（$P<0.05$），但是与对照组相比并无显著性差异（$P>0.05$）。在脑中各处理组中，HO-1 在0.5h 0.5mg/L处理和6h 3mg/L处理中，表达量均显著高于对照组和0.5h 3mg/L组（$P<0.05$），其余各组与对照组相比并无显著差异，各组间也无显著差异（$P>0.05$）。*HO-2* 在所有低氧处理后，各组表达量均显著低于对照组（$P<0.05$）。鳃中 HO-1 在对照组和所有低氧处理组内均无显著差异（$P>0.05$）。*HO-2* 基因在鳃中，低氧处理24h 3mg/L时表达量最高，显著高于其他低氧处理组（除复氧组外）（$P<0.05$）。*HO-1* 和 *HO-2* 基因在心脏中，只有0.5h 3mg/L处理组显著高于对照组和其他各组（$P<0.05$），其他各组无显著性差异（$P>0.05$）。此外，FISH 结果证实了24h 3mg/L处理组中表达量高于对照组（彩图15）。

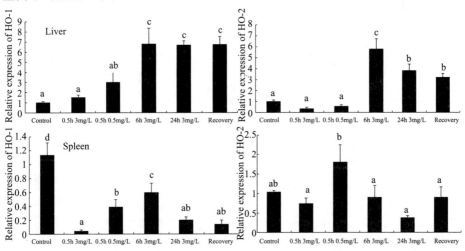

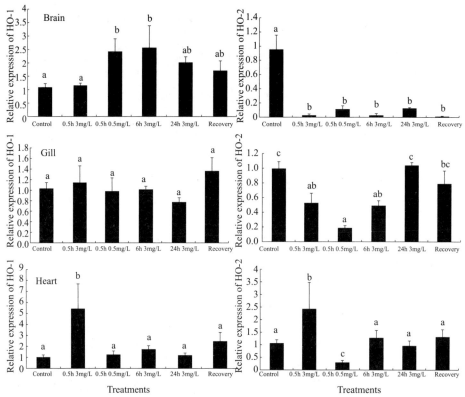

图 6-17 洛氏鲅 *HO-1* 和 *HO-2* 基因经低氧处理后的表达情况

注：不同字母代表差异显著（$P<0.05$）。

（二）*HO-1* 和 *HO-2* 基因在长期持续和昼夜低氧胁迫后的表达模式

如图 6-18 所示，*HO-1* 基因在心脏中经持续低氧处理后，呈现先上升后下降的趋势，在 14d 达到最高值，显著高于处理 7d 和 28d 时（$P<0.05$）。*HO-1* 基因经昼夜低氧处理后在 14d 达到最高值，显著高于对照组和其他低氧处理时间（$P<0.05$）。*HO-2* 基因呈现先上升后下降的趋势，在低氧处理 14d 时达到最高值，显著高于对照组和低氧处理 7d 和 21d（$P<0.05$）。经过昼夜低氧处理后，*HO-2* 基因表达量在 7d 和 14d 显著高于对照组（$P<0.05$）。在持续低氧处理的洛氏鲅脑组织中（图 6-19），*HO-1* 基因除了 7d 时与对照组差异不显著之外，其余各组均高于对照组。在 14d 和 28d 表达量显著高于对照组和处理 7d 组（$P<0.05$）。在昼夜低氧处理时，*HO-1* 基因表现为先升高后降低的模式，在昼夜低氧处理 14d 表达量最高，显著高于对照组和其他组（$P<0.05$）。*HO-2* 基因在持续低氧处理时的 28d 显著高于对照组和其他各组（$P<0.05$），为最高值。*HO-2* 基因在昼夜低氧处理后，14d 表达量最高，显著高于其他低氧处理时间组（$P<0.05$）。

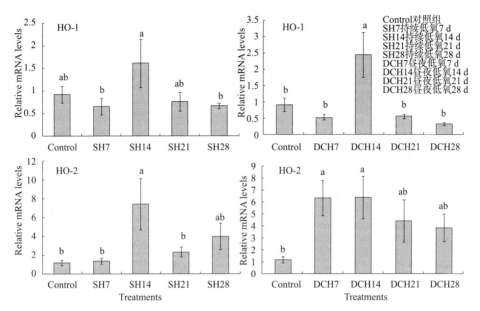

图 6-18 *HO-1* 和 *HO-2* 基因在心脏中响应低氧的表达模式

注：不同字母的组之间存在显著差异（$P<0.05$）。

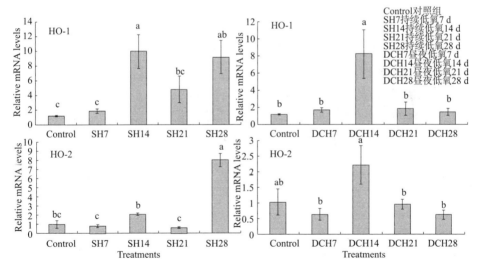

图 6-19 *HO-1* 和 *HO-2* 基因在脑中响应低氧的表达模式

注：不同字母的组之间存在显著差异（$P<0.05$）。

（三）综合分析

急性缺氧实验结果表明，HO-1 和 HO-2 在不同组织中均有显著变化。*HO-1* 和 *HO-2* 基因表达量在肝脏中均显著上调，但在缺氧期间鳃中下调或没有显著变化。肝脏是代谢变化的主要靶器官，也是重要的排泄器官，对外界刺激特别敏感（Heath，1988，Wiseman et al.，2007）。基因在鱼肝组织中的调控作用已成为热门话题。在军曹鱼中证明，肝脏的 *HO-1* 表达下降与肝组织损伤密切相关（Xie et al.，2021）。然而，在本研究中，基因表达水平显示在 6h 和 24h 缺氧和复氧肝脏中 HO 表达显著增加，表明肝脏未发生组织损伤。本研究中的一个假设是 *HO* 基因在相对长期的缺氧过程中发挥作用，因此参与了洛氏鲅的缺氧耐受性。与我们的结果相似，当暴露于极低氧气浓度时，团头鲂大脑中 *HO-2a* 表达量增加（Zhang et al.，2017）。与洛氏鲅的 HO-1 一样，团头鲂大脑中 HO-1a 在缺氧处理后显著降低（Guan et al.，2017）。与军曹鱼和鲫相比，缺氧应激后鳃中 *HO* 基因的表达迅速增加（Tzaneva and Perry，2014；Xie et al.，2021），而洛氏鲅中的 *HO-1* 表达在缺氧时没有差异。此外，在金鱼的研究中，证实 HO-1 似乎减弱了 7℃ 条件下鱼对缺氧的通气频率响应，但是不影响 25℃ 条件下鱼的通气频率响应，这表明 HO-1 在控制低氧气浓度下金鱼的呼吸作用（Tzaneva and Perry，2014）。在本研究中，鳃中 *HO-1* 表达没有显著差异，这可能与缺氧时洛氏鲅的耗氧量增加有关。当损伤发生时，*HO* 基因保护鳃细胞的生长，并参与保护鳃组织，并免于缺氧复氧损伤（Li et al.，2018；Sun et al.，2003）。多项研究表明，HO 参与心脏功能和预防心脏损伤。有研究者提出，HO-1 的强制表达减缓了自然衰老小鼠和原代培养的新生小鼠心肌细胞中的心肌细胞衰老，揭示了 HO-1 改善了心脏功能（Shan et al.，2019）。有报道提出，HO-1 通过调节线粒体质量控制在保护心脏免受氧化损伤方面发挥作用（Hull et al.，2016）。本研究中研究的 *HO-1* 和 *HO-2* 表达水平在 0.5h 缺氧处理后有所增加，这可能是由于氧气在急性缺氧期间保护心脏免受损伤，但在相对长期的缺氧应激下恢复正常。同时，我们的 FISH 结果表明，在长期缺氧应激下，*HO* 基因在肝脏中高表达。

参 考 文 献

陈秀梅，2018. L-肉碱对洛氏鱥幼鱼氧化应激的保护作用及其机理研究 [D]. 长春：吉林农业大学.
程湘军，2010. 野生柳根鱼池塘人工驯养技术 [J]. 黑龙江水产，1：20-22.
杜业峰，张爽，2019. 黑龙江省柳根鱼养殖现状及未来展望 [J]. 黑龙江水产，5：9-12.
郭文学，张永泉，佟广香，等，2015. 黑龙江流域绥芬河水系野生洛氏鱥胚胎发育 [J]. 生态学杂志，34（9）：2530-2536.
康鑫，张远，张楠，等，2011. 太子河洛氏鱥幼鱼栖息地适宜度评估 [J]. 生态毒理学报，3（6）：310-320.
孔令杰，付鹏，刘凤志，等，2018. 黑龙江省柳根鱼养殖现状及发展对策 [J]. 黑龙江水产，4：3-5.
瞿子惠，吴莉芳，周错，等，2019. 饲料碳水化合物水平对洛氏鱥消化酶和糖代谢酶活性的影响 [J]. 西北农林科技大学学报（自然科学版），2（47）：25-32.
王春清，吕树臣，何玉华，等，2014. 东北地区柳根鱼的池塘驯养技术 [J]. 饲料工业，35（6）：27-28.
王茂林，李保民，姜玉声，等，2013. 本溪太子河流域洛氏鱥个体繁殖力研究 [J]. 长江大学学报（自科版），10（29）：28-32，26-27.
王岐山，1960. 东北的鱥鱼 [J]. 动物学杂志（1）：32-34.
杨兰，吴莉芳，瞿子惠，等，2018. 饲料蛋白质水平对洛氏鱥生长、非特异性免疫及蛋白质合成的影响 [J]. 水生生物学报，42（4）：709-718.
张永泉，白庆利，郭文学，等，2013. 洛氏鱥形态性状对体质量的影响 [J]. 生态学杂志，32（11）：3063-3068.
张永泉，白庆利，徐伟，等，2015. 黑龙江流域绥芬河水系洛氏鱥个体繁殖力的研究 [J]. J 水产学杂志，28（1）：29-33.
张永泉，尹家胜，杜佳，等，2013. 雌雄洛氏鱥肌肉营养成分的比较分析 [J]. 食品科学，17：277-280.
张永泉，尹家胜，马波，等，2015. 温度和流速对洛氏鱥呼吸代谢的影响 [J]. 生态学报，35（17）：5606-5611.
赵文阁，2018. 黑龙江省鱼类原色图鉴 [M]. 北京：科学出版社.
周错，2018. 饲料脂肪水平对洛氏鱥生长、抗氧化能力及脂肪酸组成的影响 [D]. 长春：吉林农业大学.
邹瑞兴，闫磊，祖岫杰，等，2015. 洛氏鱥的研究现状及养殖发展前景 [J]. 吉林畜牧兽医，36（11）：19-22.

Abdel-Tawwab M, Monier M N, Hoseinifar S H, et al, 2019. Fish response to hypoxia stress: growth, physiological, and immunological biomarkers [J]. Fish Physiology and Biochemistry, 45 (3): 997-1013.

Aceto A, Amicarelli F, Sacchetta P, et al, 1994. Developmental aspects of detoxifying enzymes in fish (*Salmo iridaeus*) [J]. Free Radical Research, 21 (5): 285-294.

Adamovich Y, Ladeuix B, Golik M, et al, 2016. Rhythmic oxygen levels reset circadian clocks through HIF1α [J]. Cell Metabolism, 25 (1): 93-101.

Affonso E G, Polez V L P, Corrêa C F, et al, 2002. Blood parameters and metabolites in the teleost fish *Colossoma macropomum* exposed to sulfide or hypoxia [J]. Comparative Biochemistry and Physiology Part C: Toxicology & Pharmacology, 133 (3): 375-382.

Al-Bairuty G A, Shaw B J, Handy R D, et al, 2013. Histopathological effects of waterborne copper nanoparticles and copper sulphate on the organs of rainbow trout (*Oncorhynchus mykiss*) [J]. Aquat Toxicol, 126: 104-115.

Andreeva A, Skverchinskaya E, Gambaryan S, et al, 2018. Hypoxia inhibits the regulatory volume decrease in red blood cells of common frog (*Rana temporaria*) [J]. Comparative Biochemistry and Physiology Part A: Molecular & Integrative Physiology, 219-220: 44-47.

Aruoja V, Dubourguier H-C, Kasemets K, et al, 2009. Toxicity of nanoparticles of CuO, ZnO and TiO_2 to microalgae *Pseudokirchneriella subcapitata* [J]. Science of The Total Environment, 407 (4): 1461-1468.

Augustus A S, Kako Y, Yagyu H, et al, 2003. Routes of FA delivery to cardiac muscle: modulation of lipoprotein lipolysis alters uptake of TG-derived FA [J]. American Journal of Physiology-Endocrinology and Metabolism, 284 (2): E331-E339.

Avila-Flores R, Medellin R A, 2004. Ecological, taxonomic, and physiological correlates of cave use by mexican bats [J]. Journal of Mammalogy, 85 (4): 675-687.

Bailey D, Davies B, Milledge J, et al, 2000. Elevated plasma cholecystokinin at high altitude: metabolic implications for the anorexia of acute mountain sickness [J]. High altitude medicine & biology, 1: 9-23.

Bao J, Li X, Xing Y, et al, 2020. Effects of hypoxia on immune responses and carbohydrate metabolism in the Chinese mitten crab, *Eriocheir sinensis* [J]. Aquaculture Research, 51 (7): 2735-2744.

Bao J, Qiang J, Tao Y, et al, 2018. Responses of blood biochemistry, fatty acid composition and expression of microRNAs to heat stress in genetically improved farmed tilapia (*Oreochromis niloticus*) [J]. Journal of Thermal Biology, 73: 91-97.

Baptista R B, Souza-Castro N, Almeida-Val V M F, 2016. Acute hypoxia up-regulates HIF-1α and VEGF mRNA levels in Amazon hypoxia-tolerant Oscar (*Astronotus ocellatus*) [J]. Fish Physiology and Biochemistry, 42 (5): 1307-1318.

Barica J, Mathias J A, 1979. Oxygen depletion and winterkill risk in small prairie lakes under extended ice cover [J]. Journal of the Fisheries Research Board of Canada, 36 (8): 980-986.

参考文献

Barnes J, Tischkau S, Barnes J, et al, 2003. Requirement of mammalian timeless for circadian rhythmicity [J]. Science, 302: 439-442.

Barnes R, King H, Carter C G, 2011. Hypoxia tolerance and oxygen regulation in Atlantic salmon, *Salmo salar* from a Tasmanian population [J]. Aquaculture, 318 (3): 397-401.

Bensaad K, Tsuruta A, Selak M A, et al, 2006. TIGAR, a p53-Inducible Regulator of Glycolysis and Apoptosis [J]. Cell, 126 (1): 107-120.

Bensaad K, Vousden K H, 2007. p53: new roles in metabolism [J]. Trends Cell Biol, 17 (6): 286-291.

Benson B B, Krause Jr D, 1984. The concentration and isotopic fractionation of oxygen dissolved in freshwater and seawater in equilibrium with the atmosphere1 [J]. Limnology and Oceanography, 29 (3): 620-632.

Betancor M, Sprague M, Ortega A, et al, 2020. Central and peripheral clocks in Atlantic bluefin tuna (*Thunnus thynnus*, L.): Daily rhythmicity of hepatic lipid metabolism and digestive genes [J]. Aquaculture, 523: 735220.

Borowiec B G, Darcy K L, Gillette D M, et al, 2015. Distinct physiological strategies are used to cope with constant hypoxia and intermittent hypoxia in killifish (*Fundulus heteroclitus*) [J]. The Journal of Experimental Biology, 218 (8): 1198.

Bottalico L N, Weljie A M, 2021. Cross-species physiological interactions of endocrine disrupting chemicals with the circadian clock [J]. General and Comparative Endocrinology, 301 (1-2): 113650.

Boyd C, Tucker C, 1998. Pond aquaculture water quality management [M]. Massachusetts: Kluwer Academic Publishers.

Boyko M, Melamed I, Gruenbaum B F, et al, 2012. The effect of blood glutamate scavengers oxaloacetate and pyruvate on neurological outcome in a rat model of *Subarachnoid hemorrhage* [J]. Neurotherapeutics, 9 (3): 649-657.

Brady D C, Targett T E, Tuzzolino D M, 2009. Behavioral responses of juvenile weakfish (*Cynoscion regalis*) to diel-cycling hypoxia: swimming speed, angular correlation, expected displacement, and effects of hypoxia acclimation [J]. Canadian Journal of Fisheries and Aquatic Sciences, 66 (3): 415-424.

Brauner C, Ballantyne C, Randall D, et al, 2011. Air breathing in the armoured catfish (*Hoplosternum littorale*) as an adaptation to hypoxic, acidic, and hydrogen sulphide rich waters [J]. Canadian Journal of Zoology, 73 (4): 739-744.

Breitburg D, Levin L, Oschlies A, et al, 2018. Declining oxygen in the global ocean and coastal waters [J]. Science, 359.

Broeg K, 2003. Acid phosphatase activity in liver macrophage aggregates as a marker for pollution-induced immunomodulation of the non-specific immune response in fish [J]. Helgoland marine research, 57 (3-4): 166-175.

Bruick R K, 2003. Oxygen sensing in the hypoxic response pathway: regulation of the hypoxia-inducible transcription factor [J]. Genes & development, 17 (21): 2614 - 2623.

Cachot J, Galgani F, Vincent F, 1998. cDNA cloning and expression analysis of flounder p53 tumour suppressor gene [J]. Comparative Biochemistry and Physiology Part B: Biochemistry and Molecular Biology, 121 (3): 235 - 242.

Caelles C, Helmberg A, Karin M, 1994. p53-dependent apoptosis in the absence of transcriptional activation of p53-target genes [J]. Nature, 370 (6486): 220 - 223.

Cahill G M, 2002. Clock mechanisms in zebrafish [J]. Cell and tissue research, 309 (1): 27 - 34.

Cai X, Huang Y, Zhang X, et al, 2014. Cloning, characterization, hypoxia and heat shock response of hypoxia inducible factor-1 (HIF-1) from the small abalone *Haliotis diversicolor* [J]. Gene, 534 (2): 256 - 264.

Cao Y, Murphy K, McIntyre T, et al, 2000. Expression of fatty acid-CoA ligase 4 during development and in brain [J]. FEBS Letters, 467: 263 - 267.

Castro J S, Braz-Mota S, Campos D F, et al, 2020. High temperature, ph, and hypoxia cause oxidative stress and impair the spermatic performance of the amazon fish *Colossoma macropomum* [J]. Frontiers in physiology, 11: 772 - 772.

Chapman L J, McKenzie D J, 2009. Behavioral Responses and Ecological Consequences [M]. London: Elsevier.

Chapman L, Chapman C, Nordlie F, et al, 2002. Physiological refugia: Swamps, hypoxia tolerance and maintenance of fish diversity in the Lake Victoria region [J]. Comparative biochemistry and physiology Part A: Molecular & integrative physiology, 133: 421 - 437.

Chen D, Li M, Luo J, et al, 2003. Direct interactions between and Mdm2 modulate p53 function [J]. Journal of Biological Chemistry, 278 (16): 13595 - 13598.

Chen N, Chen L P, Zhang J, et al, 2012. Molecular characterization and expression analysis of three hypoxia-inducible factor alpha subunits, HIF-1α/2α/3α of the hypoxia-sensitive freshwater species, Chinese sucker [J]. Gene, 498 (1): 81 - 90.

Chilov D, Hofer T, Bauer C, et al, 2002. Hypoxia affects expression of circadian genes PER1 and CLOCK in mouse brain [J]. FASEB journal: official publication of the Federation of American Societies for Experimental Biology, 15: 2613 - 2622.

Chirala S, Chang H, Matzuk M, et al, 2003. Fatty acid synthesis is essential in embryonic development: Fatty acid synthase null mutants and most of the heterozygotes die in utero [J]. Proceedings of the National Academy of Sciences of the United States of America, 100: 6358 - 6363.

Collins G M, Clark T D, Carton A G, 2016. Physiological plasticity v. inter-population variability: Understanding drivers of hypoxia tolerance in a tropical estuarine fish [J]. Marine and Freshwater Research, 67 (10): 1575 - 1582.

Connett R J, Honig C R, Gayeski T E, et al, 1990. Defining hypoxia: a systems view of VO_2, glycolysis, energetics, and intracellular PO_2 [J]. Journal of Applied Physiology,

68 (3): 833-842.

Connor K, Gracey A, 2011. Circadian cycles are the dominant transcriptional rhythm in the intertidal mussel *Mytilus californianus* [J]. Proceedings of the National Academy of Sciences of the United States of America, 108: 16110-16115.

Cooper R U, Clough L M, Farwell M A, et al, 2002. Hypoxia-induced metabolic and antioxidant enzymatic activities in the estuarine fish *Leiostomus xanthurus* [J]. Journal of Experimental Marine Biology and Ecology, 279 (1): 1-20.

Cossins A, Gibson J S, 1997. Volume-sensitive transport systems and Volume homeostasis in vertebrate red blood cells [J]. The Journal of Experimental Biology, 200: 343-352.

Cummings D, Overduin J, 2007. Gastrointestinal Regulation of Food Intake [J]. The Journal of clinical investigation, 117: 13-23.

Cypher A D, Ickes J R, Bagatto B, 2015. Bisphenol A alters the cardiovascular response to hypoxia in *Danio rerio* embryos [J]. Comparative Biochemistry and Physiology Part C: Toxicology & Pharmacology, 174-175: 39-45.

D'Avanzo C, Kremer J N, 1994. Diel oxygen dynamics and anoxic events in an eutrophic estuary of Waquoit Bay, Massachusetts [J]. Estuaries, 17 (1): 131-139.

Dan X, Yan G, Zhang A, et al, 2014. Effects of stable and diel-cycling hypoxia on hypoxia tolerance, postprandial metabolic response, and growth performance in juvenile qingbo (*Spinibarbus sinensis*) [J]. Aquaculture, 428-429: 21-28.

Dawood M A O, Eweedah N M, Moustafa E M, et al, 2020. Probiotic effects of Aspergillus oryzae on the oxidative status, heat shock protein, and immune related gene expression of Nile tilapia (*Oreochromis niloticus*) under hypoxia challenge [J]. Aquaculture, 520: 734669.

Dawson N J, Storey K B, 2012. An enzymatic bridge between carbohydrate and amino acid metabolism: regulation of glutamate dehydrogenase by reversible phosphorylation in a severe hypoxia-tolerant crayfish [J]. Journal of Comparative Physiology B, 182 (3): 331-340.

De Boeck G, Wood C M, Iftikar F I, et al, 2013. Interactions between hypoxia tolerance and food deprivation in Amazonian oscars, *Astronotus ocellatus* [J]. The Journal of Experimental Biology, 216 (24): 4590.

de Fromentel C C, Pakdel F, Chapus A, et al, 1992. Rainbow trout p53: cDNA cloning and biochemical characterization [J]. Gene, 112 (2): 241-245.

Dhillon R, Yao L, Matey V, et al, 2013. Interspecific differences in hypoxia-induced gill remodeling in carp [J]. Physiological and Biochemical Zoology, 86: 727-739.

Diaz A, González-Castro M, García A, et al, 2008. Gross morphology and surface ultrastructure of the gills of *Odontesthes argentinensis* (Actinopterygii, Atherinopsidae) from a Southwestern Atlantic coastal lagoon [J]. Tissue & cell, 41: 193-198.

Diaz M, Vraskou Y, Gutierrez J, et al, 2009. Expression of rainbow trout glucose transporters GLUT1 and GLUT4 during in vitro muscle cell differentiation and regulation by insulin and IGF-I [J]. Am J Physiol Regul Integr Comp Physiol, 296 (3): R794-800.

Diaz R, Breitburg D, 2009. Chapter 1 The Hypoxic Environment [J]. Fish Physiology, 27: 1-23.

Ding J, Liu C, Luo S, et al, 2019. Transcriptome and physiology analysis identify key metabolic changes in the liver of the large yellow croaker (*Larimichthys crocea*) in response to acute hypoxia [J]. Ecotoxicology and Environmental Safety, 189: 109957.

Dolci G S, Rosa H Z, Vey L T, et al, 2017. Could hypoxia acclimation cause morphological changes and protect against Mn-induced oxidative injuries in silver catfish (*Rhamdia quelen*) even after reoxygenation? [J]. Environmental Pollution, 224: 466-475.

Doll S, Proneth B, Tyurina Y, et al, 2016. ACSL4 dictates ferroptosis sensitivity by shaping cellular lipid composition [J]. Nature chemical biology, 13 (1): 91-98.

Douxfils J, Deprez M, Mandiki S N M, et al, 2012. Physiological and proteomic responses to single and repeated hypoxia in juvenile Eurasian perch under domestication-Clues to physiological acclimation and humoral immune modulations [J]. Fish & Shellfish Immunology, 33 (5): 1112-1122.

Eddy S F, Morin P, Storey K B, 2005. Cloning and expression of PPARγ and PGC-1α from the hibernating ground squirrel, *Spermophilus tridecemlineatus* [J]. Molecular and Cellular Biochemistry, 269 (1): 175-182.

Edgar R S, Green E W, Zhao Y, et al, 2012. Peroxiredoxins are conserved markers of circadian rhythms [J]. Nature, 485 (7399): 459-464.

Egg M, Köblitz L, Hirayama J, et al, 2013. Linking oxygen to time: the bidirectional interaction between the hypoxic signaling pathway and the circadian clock [J]. Chronobiology International, 30: 510-529.

Ekambaram P, Narayanan M, Parasuraman P, 2017. Differential expression of survival proteins during decreased intracellular oxygen tension in brain endothelial cells of grey mullets [J]. Marine Pollution Bulletin, 115 (1): 421-428.

Evans D, Piermarini P, Choe K, 2005. The multifunctional fish gill: dominant site of gas exchange, osmoregulation, acid-base regulation, and excretion of nitrogenous waste [J]. Physiological reviews, 85: 97-177.

Everett M V, Antal C E, Crawford D L, 2012. The effect of short-term hypoxic exposure on metabolic gene expression [J]. Journal of Experimental Zoology Part A: Ecological Genetics and Physiology, 317A (1): 9-23.

Fandrey J, Gorr T A, Gassmann M, 2006. Regulating cellular oxygen sensing by hydroxylation [J]. Cardiovascular Research, 71 (4): 642-651.

Farcy E, Fleury C, Lelong C, et al, 2008. Molecular cloning of a new member of the p53 family from the Pacific oyster *Crassostrea gigas* and seasonal pattern of its transcriptional expression level [J]. Marine Environmental Research, 66 (2): 300-308.

Farhat E, Devereaux M E M, Pamenter M E, et al, 2020. Naked mole-rats suppress energy metabolism and modulate membrane cholesterol in chronic hypoxia [J]. American Journal of Physiology-Regulatory, Integrative and Comparative Physiology, 319 (2): R148-R155.

Farrell A, Richards J, 2009. Chapter 11 defining hypoxia: an integrative synthesis of the responses of fish to hypoxia [J]. Fish Physiology, 27: 487-503.

Fazio F, 2019. Fish hematology analysis as an important tool of aquaculture: A review [J]. Aquaculture, 500: 237-242.

Felix-Portillo M, Martínez-Quintana J A, Arenas-Padilla M, et al, 2016. Hypoxia drives apoptosis independently of p53 and metallothionein transcript levels in hemocytes of the whiteleg shrimp *Litopenaeus vannamei* [J]. Chemosphere, 161: 454-462.

Ficke A D, Myrick C A, Hansen L J, 2007. Potential impacts of global climate change on freshwater fisheries [J]. Reviews in Fish Biology and Fisheries, 17 (4): 581-613.

Forni D, Pozzoli U, Cagliani R, et al, 2014. Genetic adaptation of the human circadian clock to day-length latitudinal variations and relevance for affective disorders [J]. Genome Biology, 15: 499.

Franco R, Sánchez-Olea R, Reyes-Reyes E M, et al, 2009. Environmental toxicity, oxidative stress and apoptosis: Ménage à Trois [J]. Mutation Research/Genetic Toxicology and Environmental Mutagenesis, 674 (1): 3-22.

Fritsche R, Nilsson S, 1993. Cardiovascular and ventilatory control during hypoxia [M]. Dordrecht: Springer Netherlands.

Fu S J, Fu C, Yan G J, et al, 2014. Interspecific variation in hypoxia tolerance, swimming performance and plasticity in cyprinids that prefer different habitats [J]. The Journal of Experimental Biology, 217 (4): 590.

Fu S, Xie X, Cao Z, 2005. Effect of dietary composition on specific dynamic action in southern catfish *Silurus meridionalis* Chen [J]. Aquaculture Research, 36: 1384-1390.

Galic N, Hawkins T, Forbes V E, 2019. Adverse impacts of hypoxia on aquatic invertebrates: A meta-analysis [J]. Science of The Total Environment, 652: 736-743.

Ganguli G, Back J, Sengupta S, et al, 2002. The p53 tumour suppressor inhibits glucocorticoid-induced proliferation of erythroid progenitors [J]. EMBO reports, 3 (6): 569-574.

Garduño Paz M V, Méndez Sánchez J F, Burggren W, et al, 2020. Metabolic rate and hypoxia tolerance in *Girardinichthys multiradiatus* (Pisces: Goodeidae), an endemic fish at high altitude in tropical Mexico [J]. Comparative Biochemistry and Physiology Part A: Molecular & Integrative Physiology, 239: 110576.

Geng X, Feng J, Liu S, et al, 2014. Transcriptional regulation of hypoxia inducible factors alpha (HIF-α) and their inhibiting factor (FIH-1) of channel catfish (*Ictalurus punctatus*) under hypoxia [J]. Comparative Biochemistry and Physiology Part B: Biochemistry and Molecular Biology, 169: 38-50.

Gilbert D, Sundby B, Gobeil C, et al, 2005. A seventy-two-year record of diminishing deep-water oxygen in the St. Lawrence estuary: The northwest Atlantic connection [J].

Limnology and Oceanography, 50 (5): 1654 - 1666.

Gilbert-Kawai E T, Milledge J S, Grocott M P W, et al, 2014. King of the mountains: tibetan and sherpa physiological adaptations for life at high altitude [J]. Physiology, 29 (6): 388 - 402.

Gonzalez R, McDonald D, 1994. The relationship between oxygen uptake and ion loss in fish from diverse habitats [J]. The Journal of Experimental Biology, 190: 95 - 108.

Gorr T, Gerst D, Hu J, et al, 2010. Hypoxia tolerance in animals: biology and application [J]. Physiological and Biochemical Zoology, 83: 733 - 752.

Góth L, 1991. A simple method for determination of serum catalase activity and revision of reference range [J]. Clinica Chimica Acta, 196 (2): 143 - 151.

Gottlieb M, Wang Y, Teichberg V, 2003. Blood-mediated scavenging of cerebrospinal fluid glutamate [J]. Journal of neurochemistry, 87: 119 - 126.

Gracey A, Lee T, Higashi R, et al, 2011. Hypoxia-induced mobilization of stored triglycerides in the euryoxic goby *Gillichthys mirabilis* [J]. The Journal of Experimental Biology, 214: 3005 - 3012.

Gracey A, Troll J, Somero G, 2001. Hypoxia-induced expression profiling in the euryoxic fish *Gillichthys mirabilis*. [J]. Proceedings of the National Academy of Sciences of the United States of America, 98: 1993 - 1998.

Gradin K, Takasaki C, Fujii-Kuriyama Y, et al, 2002. The transcriptional activation function of the HIF-like factor requires phosphorylation at a conserved threonine [J]. The Journal of biological chemistry, 277 (26): 23508 - 23514.

Graham J, 1990. Ecological, evolutionary, and physical factors influencing aquatic animal respiration [J]. Integrative and comparative biology, 30: 137 - 146.

Grimaldi B, Bellet M M, Katada S, et al, 2010. Per2 controls lipid metabolism by direct regulation of PPARγ [J]. Cell Metabolism, 12 (5): 509 - 520.

Gu Y-Z, Moran S M, Hogenesch J B, et al, 1998. Molecular characterization and chromosomal localization of a third α-class hypoxia inducible factor subunit, HIF3α [J]. Gene Expression, 7 (3): 205 - 213.

Gu Z T, Li L, Wu F, et al, 2015. Heat stress induced apoptosis is triggered by transcription-independent p53, Ca^{2+} dyshomeostasis and the subsequent Bax mitochondrial translocation [J]. Scientific Reports, 5 (1): 11497.

Guan W-Z, Guo D-D, Sun Y-W, et al, 2017. Characterization of duplicated heme oxygenase-1 genes and their responses to hypoxic stress in blunt snout bream (*Megalobrama amblycephala*) [J]. Fish Physiology and Biochemistry, 43 (2): 641 - 651.

Guo H, Fu X, Li N, et al, 2016. Molecular characterization and expression pattern of tumor suppressor protein p53 in mandarin fish, *Siniperca chuatsi* following virus challenge [J]. Fish & Shellfish Immunology, 51: 392 - 400.

Guo H, Fu X, Lin Q, et al, 2017. Mandarin fish p53: Genomic structure, alternatively spliced variant and its mRNA expression after virus challenge [J]. Fish & Shellfish Immunology, 70: 536-544.

Hägerhäll C, 1997. Succinate: Quinone oxidoreductases-Variations on a conserved theme [J]. Biochimica et biophysica acta, 1320: 107-141.

Harayama T, Riezman H, 2018. Understanding the diversity of membrane lipid composition [J]. Nature Reviews Molecular Cell Biology, 19 (5): 281-296.

Hardy K M, Follett C R, Burnett L E, et al, 2012. Gene transcripts encoding hypoxia-inducible factor (HIF) exhibit tissue-and muscle fiber type-dependent responses to hypoxia and hypercapnic hypoxia in the Atlantic blue crab, *Callinectes sapidus* [J]. Comparative Biochemistry and Physiology Part A: Molecular & Integrative Physiology, 163 (1): 137-146.

He J, Yu Y, Qin X, et al, 2019. Identification and functional analysis of the Mandarin fish (*Siniperca chuatsi*) hypoxia-inducible factor-1α involved in the immune response [J]. Fish & Shellfish Immunology, 92: 141-150.

He W, Cao Z, Fu S, 2014. Effect of temperature on hypoxia tolerance and its underlying biochemical mechanism in two juvenile cyprinids exhibiting distinct hypoxia sensitivities [J]. Comparative biochemistry and physiology. Part A, Molecular & integrative physiology, 187: 232-241.

He Z, Deng F, Ma Z, et al, 2021. Molecular characterization, expression, and H_2O_2 induction of p53 and mdm2 in the ricefield eel, Monopterus albus [J]. Aquaculture Reports, 20: 100675.

Heath A G, 1988. Anaerobic and aerobic energy metabolism in brain and liver tissue from rainbow trout (*Salmo gairdneri*) and bullhead catfish (*Ictalurus nebulosus*) [J]. Journal of Experimental Zoology, 248 (2): 140-146.

Heikkilä M, Pasanen A, Kivirikko K, et al, 2011. Roles of the human hypoxia-inducible factor (HIF)-3α variants in the hypoxia response [J]. Cellular and molecular life sciences: CMLS, 68: 3885-3901.

Heinrichs-Caldas W, Campos D, Nazaré P-S, et al, 2018. Oxygen dependent distinct expression of hif-1α gene in aerobic and anaerobic tissues of the Amazon Oscar, *Astronotus crassipinnis* [J]. Comparative Biochemistry and Physiology Part B: Biochemistry and Molecular Biology, 227: 31-38.

Hellweger F L, Jabbur M L, Johnson C H, et al, 2020. Circadian clock helps cyanobacteria manage energy in coastal and high latitude ocean [J]. The ISME Journal, 14 (2): 560-568.

Herbert N A, Steffensen J F, 2005. The response of Atlantic cod, *Gadus morhua*, to progressive hypoxia: fish swimming speed and physiological stress [J]. Marine Biology, 147 (6): 1403-1412.

Hermes-Lima M, Zenteno-Savín T, 2002. Animal response to drastic changes in oxygen availability and physiological oxidative stress [J]. Comparative Biochemistry and Physiology

Part C: Toxicology & Pharmacology, 133 (4): 537 - 556.

Heshmati J, Golab F, Morvaridzadeh M, et al, 2020. The effects of curcumin supplementation on oxidative stress, Sirtuin-1 and peroxisome proliferator activated receptor γ coactivator 1α gene expression in polycystic ovarian syndrome (PCOS) patients: A randomized placebo-controlled clinical trial [J]. Diabetes & Metabolic Syndrome: Clinical Research & Reviews, 14 (2): 77 - 82.

Holmquist-Mengelbier L, Fredlund E, Löfstedt T, et al, 2006. Recruitment of HIF-1α and HIF-2α to common target genes is differentially regulated in neuroblastoma: HIF-2α promotes an aggressive phenotype [J]. Cancer Cell, 10 (5): 413 - 423.

Hou Z S, Wen H S, Li J F, et al, 2020. Environmental hypoxia causes growth retardation, osteoclast differentiation and calcium dyshomeostasis in juvenile rainbow trout (*Oncorhynchus mykiss*) [J]. The Science of the total environment, 705: 135272.

Hu C, Wang L, Chodosh L A, et al, 2003. Differential roles of hypoxia-inducible factor 1alpha (HIF-1alpha) and HIF-2alpha in hypoxic gene regulation [J]. Molecular and Cellular Biology, 23 (24): 9361 - 9374.

Hu C-J, Iyer S, Sataur A, et al, 2006. Differential regulation of the transcriptional activities of hypoxia-inducible factor 1 alpha (hif-1α) and hif-2α in stem cells [J]. Molecular and Cellular Biology, 26 (9): 3514.

Huang C-Y, Lin H-C, Lin C-H, 2015. Effects of hypoxia on ionic regulation, glycogen utilization and antioxidative ability in the gills and liver of the aquatic air-breathing fish Trichogaster microlepis [J]. Comparative Biochemistry and Physiology Part A: Molecular & Integrative Physiology, 179: 25 - 34.

Huang J, Guo Z, Zhang J, et al, 2021. Effects of hypoxia-reoxygenation conditions on serum chemistry indicators and gill and liver tissues of cobia (*Rachycentron canadum*) [J]. Aquaculture Reports, 20: 100692.

Huang Q, Xie D, Mao H, et al, 2017. *Ctenopharyngodon idella* p53 mediates between NF-κB and PKR at the transcriptional level [J]. Fish & Shellfish Immunology, 69: 258 - 264.

Huang Z, Liu X, Ma A, et al, 2020. Molecular cloning, characterization and expression analysis of p53 from turbot *Scophthalmus maximus* and its response to thermal stress [J]. Journal of Thermal Biology, 90: 102560.

Hughes G, 1966. The dimensions of fish gills in relation to their function [J]. The Journal of Experimental Biology, 45: 177 - 195.

Hull T D, Boddu R, Guo L, et al, 2016. Heme oxygenase-1 regulates mitochondrial quality control in the heart [J]. JCI insight, 1 (2): e85817 - e85817.

Hvas M, Oppedal F, 2019. Physiological responses of farmed Atlantic salmon and two cohabitant species of cleaner fish to progressive hypoxia [J]. Aquaculture, 512: 734353.

Imaizumi T, 2010. Arabidopsis circadian clock and photoperiodism: time to think about location [J]. Current Opinion in Plant Biology, 13 (1): 83-89.

Imaoka T, Matsuda M, Mori T, 2000. Extra pituitary expression of the prolactin gene in the goldfish, African clawed frog and mouse [J]. Zoological Science (6): 791-796, 796.

Irwin M, Kaelin W, 2001. p53 Family update: p73 and p63 develop their own identities [J]. Cell growth & differentiation: the molecular biology journal of the American Association for Cancer Research, 12: 337-349.

Islam S M, Zahangir M M, Jannat R, et al, 2020. Hypoxia reduced upper thermal limits causing cellular and nuclear abnormalities of erythrocytes in Nile tilapia, Oreochromis niloticus [J]. Journal of Thermal Biology, 90: 102604.

Ivan M, Kondo K, Yang H, et al, 2001. HIFα targeted for VHL-mediated destruction by proline hydroxylation: implications for O_2 sensing [J]. Science, 292 (5516): 464-468.

Jia Y, Wang J, Gao Y, et al, 2021. Hypoxia tolerance, hematological, and biochemical response in juvenile turbot (Scophthalmus maximus. L) [J]. Aquaculture, 535: 736380.

Jiang B H, Semenza G L, Bauer C, et al, 1996. Hypoxia-inducible factor 1 levels vary exponentially over a physiologically relevant range of O_2 tension [J]. American Journal of Physiology-Cell Physiology, 271 (4): C1172-C1180.

Jimenez A, Braun E, Tobin K, 2019. How does chronic temperature exposure affect hypoxia tolerance in sheepshead minnows' (Cyprinodon variegatus) ability to tolerate oxidative stress? [J]. Fish Physiology and Biochemistry, 45 (2): 499-510.

Jin J, Yang Y, Zhu X, et al, 2018. Effects of glucose administration on glucose and lipid metabolism in two strains of gibel carp (Carassius gibelio) [J]. General and Comparative Endocrinology, 267: 18-28.

Jing G, Li Y, Xie L, et al, 2006. Metal accumulation and enzyme activities in gills and digestive gland of pearl oyster (Pinctada fucata) exposed to copper [J]. Comparative Biochemistry and Physiology Part C: Toxicology & Pharmacology, 144 (2): 184-190.

Johannsson O E, Giacomin M, Sadauskas-Henrique H, et al, 2018. Does hypoxia or different rates of re-oxygenation after hypoxia induce an oxidative stress response in Cyphocharax abramoides (Kner, 1858), a Characid fish of the Rio Negro? [J]. Comparative Biochemistry and Physiology Part A: Molecular & Integrative Physiology, 224: 53-67.

Johnston I, Bernard L, 1982. Ultrastructure and metabolism of skeletal muscle fibres in the tench: Effects of long-term acclimation to hypoxia [J]. Cell and tissue research, 227: 179-199.

Kaghad M, Bonnet H, Yang A, et al, 1997. Monoallelically expressed gene related to p53 at 1p36, a region frequently deleted in neuroblastoma and other human cancers [J]. Cell, 90 (4): 809-819.

Kajimura S, Aida K, Duan C, 2005. Insulin-like growth factor-binding protein-1 (IGFBP-1) mediates hypoxia-induced embryonic growth and developmental retardation [J]. Proceedings of the National Academy of Sciences of the United States of America, 102 (4): 1240.

Kann O, Kovács R, 2007. Mitochondria and neuronal activity [J]. American Journal of

Physiology-Cell Physiology, 292 (2): C641 - C657.

Kawabe S, Yokoyama Y, 2012. Role of hypoxia-inducible factor α in response to hypoxia and heat shock in the pacific oyster *Crassostrea gigas* [J]. Marine Biotechnology, 14 (1): 106 - 119.

Kemp P S, Tsuzaki T, Moser M L, 2009. Linking behaviour and performance: intermittent locomotion in a climbing fish [J]. Journal of Zoology, 277 (2): 171 - 178.

Kim B-M, Rhee J-S, Park G S, et al, 2011. Cu/Zn-and Mn-superoxide dismutase (SOD) from the copepod *Tigriopus japonicus*: Molecular cloning and expression in response to environmental pollutants [J]. Chemosphere, 84 (10): 1467 - 1475.

Kobayashi Y, Ishikawa-Fujiawara T, Hirayama J, et al, 2000. Molecular analysis of zebrafish photolyase/cryptochrome family: Two types of cryptochromes present in zebrafish [J]. Genes to cells: devoted to molecular & cellular mechanisms, 5: 725 - 738.

Koike N, Hida A, Numano R, et al, 1998. Identification of the mammalian homologues of the Drosophila timeless gene, Timeless1 [J]. FEBS Letters, 441 (3): 427 - 431.

Kopp R, Köblitz L, Egg M, et al, 2011. HIF signaling and overall gene expression changes during hypoxia and prolonged exercise differ considerably [J]. Physiological genomics, 43: 506 - 516.

Koumenis C, Alarcon R, Hammond E, et al, 2001. Regulation of p53 by hypoxia: dissociation of transcriptional repression and apoptosis from p53-dependent transactivation [J]. Molecular and Cellular Biology, 21 (4): 1297 - 1310.

Krause M K, Rhodes L D, Van Beneden R J, 1997. Cloning of the p53 tumor suppressor gene from the Japanese medaka (*Oryzias latipes*) and evaluation of mutational hotspots in MNNG-exposed fish [J]. Gene, 189 (1): 101 - 106.

Kwasek K, Rimoldi S, Cattaneo A, et al, 2017. The expression of hypoxia-inducible factor-1α gene is not affected by low-oxygen conditions in yellow perch (*Perca flavescens*) juveniles [J]. Fish Physiology and Biochemistry, 43: 849 - 862.

Labboun S, Tercé-Laforgue T, Roscher A, et al, 2009. Resolving the role of plant glutamate dehydrogenase. I . *in vivo* real time nuclear magnetic resonance spectroscopy experiments [J]. Plant & cell physiology, 50: 1761 - 1773.

Lai J C C, Kakuta I, Mok H O L, et al, 2006. Effects of moderate and substantial hypoxia on erythropoietin levels in rainbow trout kidney and spleen [J]. Journal of Experimental Biology, 209 (14): 2734.

Lando D, Pongratz I, Poellinger L, et al, 2000. A redox mechanism controls differential DNA binding activities of hypoxia-inducible factor (HIF) 1alpha and the HIF-like factor [J]. The Journal of biological chemistry, 275 (7): 4618 - 4627.

Lane D P, Madhumalar A, Lee A P, et al, 2011. Conservation of all three p53 family members and Mdm2 and Mdm4 in the cartilaginous fish [J]. Cell Cycle, 10 (24): 4272 - 4279.

参考文献

Lau G Y, Arndt S, Murphy M P, et al, 2019. Species-and tissue-specific differences in ROS metabolism during exposure to hypoxia and hyperoxia plus recovery in marine sculpins [J]. Journal of Experimental Biology, 222 (22): jeb206896.

Law S H W, Wu R S S, Ng P K S, et al, 2006. Cloning and expression analysis of two distinct HIF-alpha isoforms-gcHIF-1alpha and gcHIF-4alpha-from the hypoxia-tolerant grass carp, *Ctenopharyngodon idellus* [J]. BMC Molecular Biology, 7 (1): 15.

Lazado C, Kumaratunga H, Nagasawa K, et al, 2014. Daily rhythmicity of clock gene transcripts in atlantic cod fast skeletal muscle [J]. PLoS ONE, 9: e99172.

Le Goas F, May P, Ronco P, et al, 1997. cDNA cloning and immunological characterization of rabbit p53 [J]. Gene, 185 (2): 169-173.

Le P, Do H, Nyengaard J, et al, 2016. Gill remodelling and growth rate of striped catfish *Pangasianodon hypophthalmus* under impacts of hypoxia and temperature [J]. Comparative Biochemistry and Physiology Part A: Molecular & Integrative Physiology, 203: 288-296.

Lee Y-M, Rhee J-S, Hwang D-S, et al, 2008. p53 gene expression is modulated by endocrine disrupting chemicals in the hermaphroditic fish, *Kryptolebias marmoratus* [J]. Comparative Biochemistry and Physiology Part C: Toxicology & Pharmacology, 147 (2): 150-157.

Lennard R, Huddart H, 1992. The effects of hypoxic stress on the fine structure of the flounder heart (*Platichthys flesus*) [J]. Comparative Biochemistry and Physiology Part A: Physiology, 101 (4): 723-732.

Leveelahti L, Rytkönen K T, Renshaw G M C, et al, 2014. Revisiting redox-active antioxidant defenses in response to hypoxic challenge in both hypoxia-tolerant and hypoxia-sensitive fish species [J]. Fish Physiology and Biochemistry, 40 (1): 183-191.

Levine A J, 1997. p53, the cellular gatekeeper for growth and division [J]. Cell, 88 (3): 323-331.

Levine A, Finlay C, Hinds P, 2004. P53 is a tumor suppressor gene [J]. Cell, 116: 67-69.

Lewis J, Costa I, Val A, et al, 2007. Responses to hypoxia and recovery: Repayment of oxygen debt is not associated with compensatory protein synthesis in the Amazonian cichlid, *Astronotus ocellatus* [J]. The Journal of Experimental Biology, 210: 1935-1943.

Li H, Lu L, Wu M, et al, 2020. The effects of dietary extract of mulberry leaf on growth performance, hypoxia-reoxygenation stress and biochemical parameters in various organs of fish [J]. Aquaculture Reports, 18: 100494.

Li H-T, Jiang W-D, Liu Y, et al, 2017. Dietary glutamine improves the function of erythrocytes through its metabolites in juvenile carp (*Cyprinus carpio* var. Jian) [J]. Aquaculture, 474: 86-94.

Li M, Wang X, Qi C, et al, 2018. Metabolic response of Nile tilapia (*Oreochromis niloticus*) to acute and chronic hypoxia stress [J]. Aquaculture, 495: 187-195.

Li X-L, Shi H-M, Xia H-Y, et al, 2014. Seasonal hypoxia and its potential forming mechanisms in the Mirs Bay, the northern South China Sea [J]. Continental Shelf

Research, 80: 1-7.

Li Y, Ma J, Yao K, et al, 2020. Circadian rhythms and obesity: Timekeeping governs lipid metabolism [J]. Journal of Pineal Research, 69 (3): e12682.

Liu M, Tee C, Zeng F, et al, 2011. Characterization of p53 expression in rainbow trout [J]. Comparative Biochemistry and Physiology Part C: Toxicology & Pharmacology, 154 (4): 326-332.

Liu S, Jiang X, Hu X, et al, 2004. Effects of temperature on non-specific immune parameters in two scallop species: *Argopecten irradians* (Lamarck 1819) and *Chlamys farreri* (Jones & Preston 1904) [J]. Aquaculture Research, 35 (7): 678-682.

Liu X, Li K, Zhang F, et al, 2020. Ablation of glutaredoxin 1 promotes pulmonary angiogenesis and alveolar formation in hyperoxia-injured lungs by modifying HIF-1α stability and inhibiting the NF-κB pathway [J]. Biochemical and Biophysical Research Communications, 525 (2): 528-535.

Livak K J, Schmittgen T D, 2001. Analysis of relative gene expression data using real-time quantitative PCR and the $2^{-\Delta\Delta CT}$ method [J]. Methods, 25 (4): 402-408.

Lu Z, Zhan F, Yang M, et al, 2020. The immune function of heme oxygenase-1 from grass carp (*Ctenopharyngodon idellus*) in response to bacterial infection [J]. Fish & Shellfish Immunology, 112.

Lushchak V I, Bagnyukova T V, 2006. Effects of different environmental oxygen levels on free radical processes in fish [J]. Comparative Biochemistry and Physiology Part B: Biochemistry and Molecular Biology, 144 (3): 283-289.

Lushchak V, Lushchak L, Mota A, et al, 2001. Oxidative stress and antioxidant defenses in goldfish *Carassius auratus* during anoxia and reoxygenation [J]. American journal of physiology. Regulatory, integrative and comparative physiology, 280: R100-107.

Ma N, Nicholson C, Wong M, et al, 2013. Fetal and neonatal exposure to nicotine leads to augmented hepatic and circulating triglycerides in adult male offspring due to increased expression of fatty acid synthase [J]. Toxicology and Applied Pharmacology, 275 (1): 1-11.

Ma X Y, Qiang J, He J, et al, 2015. Changes in the physiological parameters, fatty acid metabolism, and SCD activity and expression in juvenile GIFT tilapia (*Oreochromis niloticus*) reared at three different temperatures [J]. Fish Physiology and Biochemistry, 41 (4): 937-950.

Macedo A K S, Santos K, Brighenti L S, et al, 2020. Histological and molecular changes in gill and liver of fish (*Astyanax lacustris* Lutken, 1875) exposed to water from the Doce basin after the rupture of a mining tailings dam in Mariana, MG, Brazil [J]. The Science of the total environment, 735: 139505.

Magoulick D, Wilzbach M, 1997. Microhabitat selection by native brook trout and introduced rainbow trout in a small pennsylvania stream [J]. Journal of Freshwater Ecology, 12: 607-614.

Mahfouz M E, Hegazi M M, El-Magd M A, et al, 2015. Metabolic and molecular responses in Nile tilapia, *Oreochromis niloticus* during short and prolonged hypoxia [J]. Marine and Freshwater Behaviour and Physiology, 48 (5): 319-340.

Mai W-J, Yan J-L, Wang L, et al, 2010. Acute acidic exposure induces p53-mediated oxidative stress and DNA damage in tilapia (*Oreochromis niloticus*) blood cells [J]. Aquatic Toxicology, 100 (3): 271-281.

Maines M D, 1997. The heme oxygenase system: a regulator of second messenger gases [J]. Annual Review of Pharmacology and Toxicology, 37 (1): 517-554.

Maines M, 2005. The heme oxygenase system: update 2005 [J]. Antioxidants & redox signaling, 7: 1761-1766.

Mandic M, Speers-Roesch B, Richards J, 2013. Hypoxia tolerance in sculpins is associated with high anaerobic enzyme activity in brain but not in liver or muscle [J]. Physiological and biochemical zoology: PBZ, 86 (1): 92-105.

Mann L I, 1970. Effects in sheep of hypoxia on levels of lactate, pyruvate, and glucose in blood of mothers and fetus [J]. Pediatric Research, 4 (1): 46-54.

Marcon J L, Wilhelm Filho D, 1999. Antioxidant processes of the wild tambaqui, *Colossoma macropomum* (Osteichthyes, Serrasalmidae) from the Amazon [J]. Comparative Biochemistry and Physiology Part C: Pharmacology, Toxicology and Endocrinology, 123 (3): 257-263.

Mariana L, Fabrizius A, Czech-Damal N, et al, 2017. Transcriptome analysis identifies key metabolic changes in the hooded seal (*Cystophora cristata*) brain in response to hypoxia and reoxygenation [J]. PLoS ONE, 12: e0169366.

Marques I J, Leito J T, Spaink H P, et al, 2008. Transcriptome analysis of the response to chronic constant hypoxia in zebrafish hearts [J]. J Comp Physiol B, 178 (1): 77-92.

Martindale J L, Holbrook N J, 2002. Cellular response to oxidative stress: Signaling for suicide and survival [J]. Journal of Cellular Physiology, 192 (1): 1-15.

Martinez A-S, Cutler C P, Wilson G D, et al, 2005. Regulation of expression of two aquaporin homologs in the intestine of the European eel: effects of seawater acclimation and cortisol treatment [J]. American Journal of Physiology-Regulatory, Integrative and Comparative Physiology, 288 (6): R1733-R1743.

Martínez M L, Landry C, Boehm R, et al, 2006. Effects of long-term hypoxia on enzymes of carbohydrate metabolism in the Gulf killifish, *Fundulus grandis* [J]. Journal of Experimental Biology, 209 (19): 3851-3861.

Mashek D, Li L, Coleman R, 2007. Long-chain acyl-CoA synthetases and fatty acid channeling [J]. Future lipidology, 2: 465-476.

Matey V, Iftikar F, De Boeck G, et al, 2011. Gill morphology and acute hypoxia: Responses of mitochondria-rich, pavement, and mucous cells in the Amazonian oscar (*Astronotus ocellatus*) and the rainbow trout (*Oncorhynchus mykiss*), two species with very different approaches to the osmo-respiratory compromise [J]. Canadian Journal of

Zoology, 89: 307 - 324.

Matey V, Richards J G, Wang Y, et al, 2008. The effect of hypoxia on gill morphology and ionoregulatory status in the Lake Qinghai scaleless carp, *Gymnocypris przewalskii* [J]. Journal of Experimental Biology, 211 (7): 1063.

Matheu A, Maraver A, Klatt P, et al, 2007. Delayed ageing through damage protection by the Arf/p53 pathway [J]. Nature, 448 (7151): 375 - 379.

Mathias J A, Barica J, 1980. Factors controlling oxygen depletion in ice-covered lakes [J]. Canadian Journal of Fisheries and Aquatic Sciences, 37 (2): 185 - 194.

Matlashewski G, Lamb P, Pim D, et al, 1984. Isolation and characterization of a human p53 cDNA clone: expression of the human p53 gene [J]. The EMBO Journal, 3 (13): 3257 - 3262.

Maynard M A, Qi H, Chung J, et al, 2003. Multiple splice variants of the human HIF-3 alpha locus are targets of the von Hippel-Lindau E3 ubiquitin ligase complex [J]. The Journal of biological chemistry, 278 (13): 11032 - 11040.

McCoubrey W K, Ewing J F, Maines M D, 1992. Human heme oxygenase-2: Characterization and expression of a full-length cDNA and evidence suggesting that the two HO-2 transcripts may differ by choice of polyadenylation signal [J]. Archives of Biochemistry and Biophysics, 295 (1): 13 - 20.

McDonald D, McMahon B, 1977. Respiratory development in Artic char *Salvelinus alpinus* under conditions of normoxia and chronic hypoxia [J]. Canadian Journal of Zoology, 55: 1461 - 1467.

McKenzie D, Martinez R, Morales Rojas A, et al, 2003. Effects of growth hormone transgenesis on metabolic rate, exercise performance and hypoxia tolerance in tilapia hybrids [J]. Journal of Fish Biology, 63: 398 - 409.

Meira L B, Bugni J M, Green S L, et al, 2008. DNA damage induced by chronic inflammation contributes to colon carcinogenesis in mice [J]. The Journal of clinical investigation, 118 (7): 2516 - 2525.

Metcalfe N B, Alonso-Alvarez C, 2010. Oxidative stress as a life-history constraint: the role of reactive oxygen species in shaping phenotypes from conception to death [J]. Functional Ecology, 24 (5): 984 - 996.

Mitrovic D, Dymowska A, Nilsson G E, et al, 2009. Physiological consequences of gill remodeling in goldfish (*Carassius auratus*) during exposure to long-term hypoxia [J]. American Journal of Physiology-Regulatory, Integrative and Comparative Physiology, 297 (1): R224 - R234.

Mitrovic D, Perry S F, 2009. The effects of thermally induced gill remodeling on ionocyte distribution and branchial chloride fluxes in goldfish (*Carassius auratus*) [J]. Journal of Experimental Biology, 212 (6): 843 - 852.

Mohindra V, Tripathi R K, Singh R K, et al, 2013. Molecular characterization and expression analysis of three hypoxia-inducible factor alpha subunits, HIF-1α, -2α and-3α in hypoxia-tolerant Indian catfish, *Clarias batrachus* (Linnaeus, 1758) [J]. Molecular Biology Reports, 40 (10): 5805-5815.

Mommsen T P, French C J, Hochachka P W, 1980. Sites and patterns of protein and amino acid utilization during the spawning migration of salmon [J]. Canadian Journal of Zoology, 58 (10): 1785-1799.

Morley S A, Peck Ls Fau-Miller A J, Miller Aj Fau-Pörtner H O, et al, Hypoxia tolerance associated with activity reduction is a key adaptation for *Laternula elliptica* seasonal energetics [J]. 153 (1): 29-36.

Moura M A F, Oliveira M I S, Val A L, 1997. Effects of hypoxia on leucocytes of two Amazon fish *Colossoma macropomum* and *Hoplosternum littorale* [J]. Revista da Universidade do Amazonas. Série Ciências Agrárias, 1: 13-22.

Mu W, Wen H, Li J, et al, 2015. HIFs genes expression and hematology indices responses to different oxygen treatments in an ovoviviparous teleost species *Sebastes schlegelii* [J]. Marine Environmental Research, 110: 142-151.

Mu Y, Li W, Wei Z, et al, 2020. Transcriptome analysis reveals molecular strategies in gills and heart of large yellow croaker (*Larimichthys crocea*) under hypoxia stress [J]. Fish & Shellfish Immunology, 104: 304-313.

Muñoz-Sánchez J, Chánez-Cárdenas M E, 2014. A review on hemeoxygenase-2: focus on cellular protection and oxygen response [J]. Oxidative Medicine and Cellular Longevity, 2014: 604981.

Mustafa S A, Karieb S S, Davies S J, et al, 2015. Assessment of oxidative damage to DNA, transcriptional expression of key genes, lipid peroxidation and histopathological changes in carp *Cyprinus carpio* L. following exposure to chronic hypoxic and subsequent recovery in normoxic conditions [J]. Mutagenesis, 30 (1): 107-116.

Mustafa S, Davies S, Jha A, 2012. Determination of hypoxia and dietary copper mediated sub lethal toxicity in carp, *Cyprinus carpio*, at different levels of biological organisation [J]. Chemosphere, 87: 413-422.

Naito A T, Okada S, Minamino T, et al, 2010. Promotion of CHIP-Mediated p53 Degradation Protects the Heart From Ischemic Injury [J]. Circulation Research, 106 (11): 1692-1702.

Nakano T, Kameda M, Shoji Y, et al, 2014. Effect of severe environmental thermal stress on redox state in salmon [J]. Redox Biology, 2: 772-776.

Nam S-E, Haque M N, Shin Y K, et al, 2020. Constant and intermittent hypoxia modulates immunity, oxidative status, and blood components of red sea bream and increases its susceptibility to the acute toxicity of red tide dinoflagellate [J]. Fish & Shellfish Immunology, 105: 286-296.

Negro C L, Collins P, 2017. Histopathological effects of chlorpyrifos on the gills, hepatopancreas and gonads of the freshwater crab *Zilchiopsis collastinensis*. Persistent effects after exposure [J]. Ecotoxicology and Environmental Safety, 140: 116-122.

Nierenberg A A, Ghaznavi S A, Sande Mathias I, et al, 2018. Peroxisome proliferator-activated receptor gamma coactivator-1 alpha as a novel target for bipolar disorder and other neuropsychiatric disorders [J]. Biological Psychiatry, 83 (9): 761-769.

Nikinmaa M, Rees B B, 2005. Oxygen-dependent gene expression in fishes [J]. American Journal of Physiology-Regulatory, Integrative and Comparative Physiology, 288 (5): R1079-R1090.

Nilsson G E, 2007. Gill remodeling in fish-a new fashion or an ancient secret? [J]. Journal of Experimental Biology, 210 (14): 2403.

Nilsson G E, Dymowska A, Stecyk J A W, 2012. New insights into the plasticity of gill structure [J]. Respiratory Physiology & Neurobiology, 184 (3): 214-222.

Nilsson G E, Renshaw G M C, 2004. Hypoxic survival strategies in two fishes: extreme anoxia tolerance in the North European crucian carp and natural hypoxic preconditioning in a coral-reef shark [J]. Journal of Experimental Biology, 207 (18): 3131.

Nilsson S, 1986. Control of gill blood flow [M]. London: Croom Helm.

Nuñez-Hernandez D M, Felix-Portillo M, Peregrino-Uriarte A B, et al, 2018. Cell cycle regulation and apoptosis mediated by p53 in response to hypoxia in hepatopancreas of the white shrimp *Litopenaeus vannamei* [J]. Chemosphere, 190: 253-259.

O'Connell E J, Martinez C A, Liang Y G, et al, 2020. Out of breath, out of time: interactions between HIF and circadian rhythms [J]. American Journal of Physiology-Cell Physiology, 319 (3): C533-C540.

Obirikorang K A, Acheampong J N, Duodu C P, et al, 2020. Growth, metabolism and respiration in Nile tilapia (*Oreochromis niloticus*) exposed to chronic or periodic hypoxia [J]. Comparative Biochemistry and Physiology Part A: Molecular & Integrative Physiology, 248: 110768.

Ohkawa H, Ohishi N, Yagi K, 1979. Assay for lipid peroxides in animal tissues by thiobarbituric acid reaction [J]. Analytical biochemistry, 95 (2): 351-358.

Oishi K, Miyazaki K, Kadota K, et al, 2003. Genome-wide expression analysis of mouse liver reveals clock-regulated circadian output genes [J]. Journal of Biological Chemistry, 278 (42): 41519-41527.

Okada Y, Maeno E, Shimizu T, et al, 2001. Receptor-mediated control of regulatory volume decrease (RVD) and apoptotic volume decrease (AVD) [J]. J Physiol, 532: 3-16.

Okamura H, 2007. Suprachiasmatic nucleus clock time in the mammalian circadian system [J]. Cold Spring Harbor symposia on quantitative biology, 72: 551-556.

Okamura Y, Mekata T, Elshopakey G E, et al, 2018. Molecular characterization and gene expression analysis of hypoxia-inducible factor and its inhibitory factors in kuruma shrimp *Marsupenaeus japonicus* [J]. Fish & Shellfish Immunology, 79: 168-174.

Oliveira G, Rossi I, Kucharski L, et al, 2004. Hepatopancreas gluconeogenesis and glycogen content during fasting in crabs previously maintained on a high-protein or carbohydrate-rich diet [J]. Comparative biochemistry and physiology Part A: Molecular & integrative physiology, 137: 383-390.

Ostadal B, Kolar F, 2007. Cardiac adaptation to chronic high-altitude hypoxia: Beneficial and adverse effects [J]. Respiratory Physiology & Neurobiology, 158 (2): 224-236.

Ou L C, Tenney S M, 1970. Properties of mitochondria from hearts of cattle acclimatized to high altitude [J]. Respiration Physiology, 8 (2): 151-159.

Paniw M, Maag N, Cozzi G, et al, 2019. Life history responses of meerkats to seasonal changes in extreme environments [J]. Science, 363: 631-635.

Park J, Tompsett A, Newsted J, et al, 2008. Fluorescence in situ hybridization techniques (FISH) to detect changes in CYP19a gene expression of Japanese medaka (*Oryzias latipes*) [J]. Toxicology and Applied Pharmacology, 232: 226-235.

Pasanen A, Heikkilä M, Rautavuoma K, et al, 2010. Hypoxia-inducible factor (HIF)-3α is subject to extensive alternative splicing in human tissues and cancer cells and is regulated by HIF-1 but not HIF-2 [J]. The international journal of biochemistry & cell biology, 42: 1189-1200.

Patton S, Zulak I M, Trams E G, 1975. Fatty acid metabolism via triglyceride in the salmon heart [J]. Journal of Molecular and Cellular Cardiology, 7 (11): 857-865.

Pelster B, Egg M, 2015. Multiplicity of hypoxia-inducible transcription factors and their connection to the circadian clock in the zebrafish [J]. Physiological and Biochemical Zoology, 88 (2): 146-157.

Pelster B, Egg M, 2018. Hypoxia-inducible transcription factors in fish: expression, function and interconnection with the circadian clock [J]. The Journal of Experimental Biology, 221 (13).

Perry SF, Esbaugh A, Braun M, et al, 2009. Gas transport and gill function in water-breathing fish [M]. Berlin: Springer-Verlag.

Petersen M, Vatner D, Shulman G, 2017. Regulation of hepatic glucose metabolism in health and disease [J]. Nature Reviews Endocrinology, 13 (10): 572-587.

Pichavant K, Maxime V, Thébault M T, et al, 2002. Effects of hypoxia and subsequent recovery on turbot *Scophthalmus maximus*: hormonal changes and anaerobic metabolism [J]. Marine Ecology Progress Series, 225: 275-285.

Pierce V A, Crawford D L, 1997. Phylogenetic Analysis of Glycolytic Enzyme Expression [J]. Science, 276 (5310): 256-259.

Pittendrigh C, 1993. Temporal organization: reflections of a darwinian clock-watcher [J]. Annu Rev Physiol, 55: 16-54.

Polakof S, Panserat S, Soengas J L, et al, 2012. Glucose metabolism in fish: a review [J]. Journal of Comparative Physiology B, 182 (8): 1015-1045.

Pollock M S, Clarke L M J, Dubé M G, 2007. The effects of hypoxia on fishes: From ecological relevance to physiological effects [J]. Environmental Reviews, 15: 1 - 14.

Pothoven S, Vanderploeg H, Ludsin S, et al, 2009. Feeding ecology of emerald shiners and rainbow smelt in central lake Erie [J]. Journal of Great Lakes Research, 35: 190 - 198.

Powell W H, Hahn M E, 2002. Identification and functional characterization of hypoxia-inducible factor 2α from the estuarine teleost, Fundulus heteroclitus: Interaction of HIF-2α with two ARNT2 splice variants [J]. Journal of Experimental Zoology, 294 (1): 17 - 29.

Pozo Devoto V M, Chavez J C, Fiszer de Plazas S, 2006. Acute hypoxia and programmed cell death in developing CNS: Differential vulnerability of chick optic tectum layers [J]. Neuroscience, 142 (3): 645 - 653.

Prokkola J, Nikinmaa M, 2018. Circadian rhythms and environmental disturbances-Underexplored interactions [J]. The Journal of Experimental Biology, 221 (16): jeb179267.

Prokkola J, Nikinmaa M, Lubiana P, et al, 2015. Hypoxia and the pharmaceutical diclofenac influence the circadian responses of three-spined stickleback [J]. Aquatic Toxicology, 158: 116 - 124.

Qi Z, Liu Y, Luo S, et al, 2013. Molecular cloning, characterization and expression analysis of tumor suppressor protein p53 from orange-spotted grouper, *Epinephelus coioides* in response to temperature stress [J]. Fish & Shellfish Immunology, 35 (5): 1466 - 1476.

Qiang J, Zhu X-W, He J, et al, 2020. Mir-34a regulates the activity of hif-1a and p53 signaling pathways by promoting glut1 in genetically improved farmed tilapia (GIFT, *Oreochromis niloticus*) under hypoxia stress [J]. Frontiers in physiology, 11: 670 - 670.

Rahman M S, Thomas P, 2007. Molecular cloning, characterization and expression of two hypoxia-inducible factor alpha subunits, HIF-1alpha and HIF-2alpha, in a hypoxia-tolerant marine teleost, Atlantic croaker (*Micropogonias undulatus*) [J]. Gene, 396 (2): 273 - 282.

Rajalakshmi S, Mohandas A, 2005. Copper-induced changes in tissue enzyme activity in a freshwater mussel [J]. Ecotoxicology and Environmental Safety, 62 (1): 140 - 143.

Rankin E B, Giaccia A J, 2008. The role of hypoxia-inducible factors in tumorigenesis [J]. Cell death and differentiation, 15: 678 - 685.

Rashid I, Baisvar V S, Singh M, et al, 2020. Isolation and characterization of hypoxia inducible heme oxygenase 1 (HMOX1) gene in *Labeo rohita* [J]. Genomics, 112 (3): 2327 - 2333.

Rey G, Reddy A, 2013. Connecting cellular metabolism to circadian clocks [J]. Trends in cell biology, 23 (5): 234 - 241.

Richards J G, 2010. Metabolic rate suppression as a mechanism for surviving environmental challenge in fish [M]. Berlin: Heidelberg.

Richards J, 2009. Chapter 10 metabolic and molecular responses of fish to hypoxia [J]. Fish

Physiology, 27: 443-485.

Richards J, 2011. Physiological, behavioral and biochemical adaptations of intertidal fishes to hypoxia [J]. The Journal of Experimental Biology, 214: 191-199.

Riesco-Fagundo A M, Perez-Garcia M T, Gonzalez C, et al, 2001. O_2 modulates large-conductance Ca^{2+}-dependent K^+ channels of rat chemoreceptor cells by a membrane-restricted and CO-sensitive mechanism [J]. Circulation Research, 89 (5): 430-436.

Rimoldi S, Terova G, Ceccuzzi P, et al, 2012. HIF-1α mRNA levels in Eurasian perch (*Perca fluviatilis*) exposed to acute and chronic hypoxia [J]. Molecular Biology Reports, 39 (4): 4009-4015.

Rimoldi S, Terova G, Zaccone G, et al, 2016. The effect of hypoxia and hyperoxia on growth and expression of hypoxia-related genes and proteins in spotted gar *Lepisosteus oculatus* larvae and juveniles [J]. Journal of Experimental Zoology Part B: Molecular and Developmental Evolution, 326 (4): 250-267.

Rissanen E, Tranberg H K, Sollid J, et al, 2006. Temperature regulates hypoxia-inducible factor-1 (HIF-1) in a poikilothermic vertebrate, crucian carp (*Carassius carassius*) [J]. Journal of Experimental Biology, 209 (6): 994-1003.

Roberts J J, Grecay P A, Ludsin S A, et al, 2012. Evidence of hypoxic foraging forays by yellow perch (*Perca flavescens*) and potential consequences for prey consumption [J]. Freshwater Biology, 57 (5): 922-937.

Robinson C, 2019. Microbial respiration, the engine of ocean deoxygenation [J]. Frontiers in Marine Science, 5.

Rocha-Santos C, Bastos F F, Dantas R F, et al, 2018. Glutathione peroxidase and glutathione S-transferase in blood and liver from a hypoxia-tolerant fish under oxygen deprivation [J]. Ecotoxicology and Environmental Safety, 163: 604-611.

Rojas D A, Perez-Munizaga D A, Centanin L, et al, 2007. Cloning of HIF-1α and HIF-2α and mRNA expression pattern during development in zebrafish [J]. Gene Expression Patterns, 7 (3): 339-345.

Rye P, Lamarr W, 2015. Measurement of glycolysis reactants by high-throughput SPE-MS/MS: characterization of ppi-dependent phosphofructokinase as a case study [J]. Analytical biochemistry, 482: 40-47.

Rytkönen K T, Vuori K A M, Primmer C R, et al, 2007. Comparison of hypoxia-inducible factor-1 alpha in hypoxia-sensitive and hypoxia-tolerant fish species [J]. Comparative Biochemistry and Physiology Part D: Genomics and Proteomics, 2 (2): 177-186.

Saroglia M, Cecchini S, Terova G, et al, 2000. Influence of environmental temperature and water oxygen concentration on gas diffusion distance in sea bass (*Dicentrarchus labrax*, L.) [J]. Fish Physiology and Biochemistry, 23: 55-58.

Saroglia M, Terova G, De Stradis A, et al, 2002. Morphometric adaptations of sea bass gills to different dissolved oxygen partial pressure [J]. Journal of Fish Biology, 60: 1423-1420.

Scandalios J G, 2005. Oxidative stress: Molecular perception and transduction of signals triggering antioxidant gene defenses [J]. Brazilian journal of medical and biological research, 38: 995-1014.

Schmidt H, Wegener G, 1988. Glycogen phosphorylase in fish brain (*Carassius carassius*) during hypoxia [J]. Biochemical Society Transactions, 16 (4): 621-622.

Scott A L, Rogers W A, 2006. Haematological effects of prolonged hypoxia on channel catfish *Ictalurus punctatus* (Rafinesque) [J]. Journal of Fish Biology, 18: 591-601.

Scot-Taggart C, Van Cauwenberghe O, D M, et al, 2002. Regulation of Γ-aminobutyric acid synthesis *in situ* by glutamate availability [J]. Physiologia Plantarum, 106: 363-369.

Semenza G L, 1998. Hypoxia-inducible factor 1: master regulator of O_2 homeostasis [J]. Current Opinion in Genetics & Development, 8 (5): 588-594.

Semenza G L, 2000. HIF-1: mediator of physiological and pathophysiological responses to hypoxia [J]. Journal of Applied Physiology, 88 (4): 1474-1480.

Serebrovska T V, Portnychenko A G, Portnichenko V I, et al, 2019. Effects of intermittent hypoxia training on leukocyte pyruvate dehydrogenase kinase 1 (PDK-1) mRNA expression and blood insulin level in prediabetes patients [J]. European Journal of Applied Physiology, 119 (3): 813-823.

Shan H, Li T, Zhang L, et al, 2019. Heme oxygenase-1 prevents heart against myocardial infarction by attenuating ischemic injury-induced cardiomyocytes senescence [J]. EBioMedicine, 39: 59-68.

Shekhawat G S, Verma K, 2010. Haem oxygenase (HO): an overlooked enzyme of plant metabolism and defence [J]. Journal of Experimental Botany, 61 (9): 2255-2270.

Shen R-J, Jiang X-Y, Pu J-W, et al, 2010. HIF-1α and-2α genes in a hypoxia-sensitive teleost species Megalobrama amblycephala: cDNA cloning, expression and different responses to hypoxia [J]. Comparative Biochemistry and Physiology Part B: Biochemistry and Molecular Biology, 157 (3): 273-280.

Shi Z, Liu K, Zhang S, et al, 2019. Spatial distributions of mesozooplankton biomass, community composition and grazing impact in association with hypoxia in the Pearl River Estuary [J]. Estuarine, Coastal and Shelf Science, 225.

Silva G, Matos L, Freitas J, et al, 2019. Gene expression, genotoxicity, and physiological responses in an Amazonian fish, *Colossoma macropomum* (Cuvier 1818), exposed to Roundup® and subsequent acute hypoxia [J]. Comparative Biochemistry and Physiology Part C: Toxicology & Pharmacology, 222.

Sionov R V, Haupt Y, 1999. The cellular response to p53: the decision between life and death [J]. Oncogene, 18 (45): 6145-6157.

Sogawa K, Fujii-Kuriyama Y, 1997. Ah receptor, a novel ligand-activated transcription factor [J]. Journal of biochemistry, 122 (6): 1075-1079.

Soitamo A J, Rabergh C M, Gassmann M, et al, 2001. Characterization of a hypoxia-inducible factor (HIF-1alpha) from rainbow trout. Accumulation of protein occurs at normal venous oxygen tension [J]. The Journal of biological chemistry, 276 (23): 19699 – 19705.

Soldatov A A, 1996. The effect of hypoxia on red blood cells of flounder: a morphologic and autoradiographic study [J]. Journal of Fish Biology, 48 (3): 321 – 328.

Sollid J, De Angelis P, Gundersen K, et al, 2003. Hypoxia induces adaptive and reversible gross morphological changes in crucian carp gills [J]. Journal of Experimental Biology, 206 (20): 3667.

Song W, Zhong C, Yuan Y, et al, 2020. Peroxisome proliferator-activated receptor-coactivator 1-beta (PGC-1β) modulates the expression of genes involved in adipogenesis during preadipocyte differentiation in chicken [J]. Gene, 741: 144516.

Soussi T, May P, 1996. Structural Aspects of the p53 Protein in Relation to Gene Evolution: A Second Look [J]. Journal of Molecular Biology, 260 (5): 623 – 637.

Spanagel R, Pendyala G, Abarca C, et al, 2005. The clock gene Per2 influences the glutamatergic system and modulates alcohol consumption [J]. Nature Medicine, 11 (1): 35 – 42.

Speers-Roesch B, Sandblom E, Lau G, et al, 2009. Effects of environmental hypoxia on cardiac energy metabolism and performance in tilapia [J]. American journal of physiology. Regulatory, integrative and comparative physiology, 298: R104 – 119.

Stefanatos R, Sanz A, 2018. The role of mitochondrial ROS in the aging brain [J]. FEBS Letters, 592 (5): 743 – 758.

Stierhoff K, Targett T, Grecay P, 2003. Hypoxia tolerance of the mummichog: The role of access to the water surface [J]. Journal of Fish Biology, 63: 580 – 592.

Stroka D M, Burkhardt T, Desbaillets I, et al, 2001. HIF-1 is expressed in normoxic tissue and displays an organ-specific regulation under systemic hypoxia [J]. FASEB journal: official publication of the Federation of American Societies for Experimental Biology, 15 (13): 2445 – 2453.

Sun J L, Liu Y F, Jiang T, et al, 2021. Golden pompano (*Trachinotus blochii*) adapts to acute hypoxic stress by altering the preferred mode of energy metabolism [J]. Aquaculture, 542: 736842.

Sun J L, Zhao L L, Wu H, et al, 2020. Acute hypoxia changes the mode of glucose and lipid utilization in the liver of the largemouth bass (*Micropterus salmoides*) [J]. Science of The Total Environment, 713.

Sun J L, Zhao L, Liao L, et al, 2019. Interactive effect of thermal and hypoxia on largemouth bass (*Micropterus salmoides*) gill and liver: Aggravation of oxidative stress, inhibition of immunity and promotion of cell apoptosis [J]. Fish & Shellfish Immunology, 98.

Sun J, Shi L, Xiao T, et al, 2021. microRNA-21, via the HIF-1α/VEGF signaling pathway, is involved in arsenite-induced hepatic fibrosis through aberrant cross-talk of hepatocytes and hepatic stellate cells [J]. Chemosphere, 266: 129177.

Sun Q, Wang D, Wei Q, 2014. The complete mitochondrial gemone of *Phoxinus lagowskii* (Teleostei, Cypriniformes: Cyprinidae) [J]. Mitochondrial DNA, 27 (2): 830-831.

Sun S, Guo Z, Fu H, et al, 2018. Integrated metabolomic and transcriptomic analysis of brain energy metabolism in the male Oriental river prawn (*Macrobrachium nipponense*) in response to hypoxia and reoxygenation [J]. Environmental Pollution, 243: 1154-1165.

Suski C D, Ridgway M S, 2009. Winter biology of centrarchid fishes [M]. West Sussex: Wiley-Blackwell.

Takeda N, Maemura K, Imai Y, et al, 2004. Endothelial pas domain protein 1 gene promotes angiogenesis through the transactivation of both vascular endothelial growth factor and its receptor, FLT-1 [J]. Circulation Research, 95 (2): 146-153.

Tenhunen R, Marver H S, Schmid R, 1968. The enzymatic conversion of heme to bilirubin by microsomal heme oxygenase [J]. Proceedings of the National Academy of Sciences of the United States of America, 61 (2): 748-755.

Terova G, Rimoldi S, Corà S, et al, 2008. Acute and chronic hypoxia affects HIF-1α mRNA levels in sea bass (*Dicentrarchus labrax*) [J]. Aquaculture, 279 (1): 150-159.

Thomé R G, de Oliveira Cardoso I C, de Oliveira S E, et al, 2018. Oogenesis is accompanied by cyclic morphological changes in hepatocytes of Neotropical freshwater fish *Piabina argentea* [J]. Anatomia, histologia, embryologia, 47 (5): 466-474.

Tort L, Balasch J, Mackenzie S, 2003. Fish immune system. A crossroads between innate and adaptive responses [J]. Inmunología, 22: 277-286.

Tosini G, Pozdeyev N, Sakamoto K, et al, 2008. The circadian clock system in the mammalian retina [J]. BioEssays, 30 (7): 624-633.

Tzaneva V, Perry S F, 2014. Heme oxygenase-1 (HO-1) mediated respiratory responses to hypoxia in the goldfish, Carassius auratus [J]. Respiratory Physiology & Neurobiology, 199: 1-8.

Van Liere E J, 1936. The effect of prolonged anoxemia on the heart and spleen in the mammal [J]. American Journal of Physiology-Legacy Content, 116 (2): 290-294.

Vanderplancke G, Claireaux G, Quazuguel P, et al, 2015. Exposure to chronic moderate hypoxia impacts physiological and developmental traits of European sea bass (*Dicentrarchus labrax*) larvae [J]. Fish Physiology and Biochemistry, 41 (1): 233-242.

Vaquer-Sunyer R, Duarte C M, 2008. Thresholds of hypoxia for marine biodiversity [J]. Proceedings of the National Academy of Sciences, 105 (40): 15452.

Vera L M, Negrini P, Zagatti C, et al, 2013. Light and feeding entrainment of the molecular circadian clock in a marine teleost (*Sparus aurata*) [J]. Chronobiology International, 30 (5): 649-661.

Volkoff H, 2015. Cloning, tissue distribution and effects of fasting on mRNA expression levels of leptin and ghrelin in red-bellied piranha (*Pygocentrus nattereri*) [J]. General and

Comparative Endocrinology, 217-218: 20-27.

Vutukuru S S, Chintada S, Radha M K, et al, 2006. Acute effects of copper on superoxide dismutase, catalase and lipid peroxidation in the freshwater teleost fish, *Esomus danricus* [J]. Fish Physiology and Biochemistry, 32 (3): 221-229.

Walshe T E, D'Amore P A, 2008. The role of hypoxia in vascular injury and repair [J]. Annual Review of Pathology: Mechanisms of Disease, 3 (1): 615-643.

Wang H, 2008. Comparative analysis of teleost fish genomes reveals preservation of different ancient clock duplicates in different fishes [J]. Marine Genomics, 1 (2): 69-78.

Wang J, Yang Y, Wang Z, et al, 2021. Comparison of effects in sustained and diel-cycling hypoxia on hypoxia tolerance, histology, physiology and expression of clock genes in high latitude fish *Phoxinus lagowskii* [J]. Comparative Biochemistry and Physiology Part A: Molecular & Integrative Physiology, 260: 111020.

Wang M, Wu F, Xie S, et al, 2021. Acute hypoxia and reoxygenation: Effect on oxidative stress and hypoxia signal transduction in the juvenile yellow catfish (*Pelteobagrus fulvidraco*) [J]. Aquaculture, 531: 735903.

Wang Q, Ao Y, Yang K, et al, 2016. Circadian clock gene Per2 plays an important role in cell proliferation, apoptosis and cell cycle progression in human oral squamous cell carcinoma [J]. Oncol Rep, 35 (6): 3387-3394.

Wang Q, Li X, Sha H, et al, 2021. Identification of microRNAs in silver carp (*Hypophthalmichthys molitrix*) response to hypoxia stress [J]. Animals, 11 (10): 2917.

Wang Q, Luo S, Ghonimy A, et al, 2019. Effect of dietary l-carnitine on growth performance and antioxidant response in Amur minnow (*Phoxinus lagowskii* Dybowskii) [J]. Aquaculture Nutrition, 25 (4): 749-760.

Wang X, Liu S, Dunham R, et al, 2017. Effects of strain and body weight on low-oxygen tolerance of channel catfish (*Ictalurus punctatus*) [J]. Aquaculture International, 25 (4): 1645-1652.

Wang X, Liu S, Jiang C, et al, 2017. Multiple across-strain and within-strain QTLs suggest highly complex genetic architecture for hypoxia tolerance in channel catfish [J]. Molecular Genetics and Genomics, 292 (1): 63-76.

Weber J-M, 2011. Metabolic fuels: Regulating fluxes to select mix [J]. The Journal of Experimental Biology, 214: 286-294.

Wells R M G, Grigg G C, Beard L A, et al, 1989. Hypoxic responses in a fish from a stable environment: blood oxygen transport in the antarctic fish *Pagothenia borchgrevinki* [J]. Journal of Experimental Biology, 141 (1): 97-111.

Wenger R H, 2002. Cellular adaptation to hypoxia: O_2-sensing protein hydroxylases, hypoxia-inducible transcription factors, and O_2-regulated gene expression [J]. The FASEB Journal, 16 (10): 1151-1162.

Wenger R H, Gassmann M, 1997. Oxygen (es) and the hypoxia-inducible factor-1 [J]. Biological chemistry, 378 (7): 609-616.

Whitehouse L M, Manzon R G, 2019. Hypoxia alters the expression of HIF-1a mRNA and downstream HIF-1 response genes in embryonic and larval lake whitefish (*Coregonus clupeaformis*) [J]. Comparative Biochemistry and Physiology Part A: Molecular & Integrative Physiology, 230: 81-90.

Whittaker G, Barnhart B, Färe R, et al, 2015. Application of index number theory to the construction of a water quality index: Aggregated nutrient loadings related to the areal extent of hypoxia in the northern Gulf of Mexico [J]. Ecological Indicators, 49: 162-168.

Williams K, Cassidy A, Verhille C, et al, 2019. Diel cycling hypoxia enhances hypoxia-tolerance in rainbow trout (*Oncorhynchus mykiss*): evidence of physiological and metabolic plasticity [J]. The Journal of Experimental Biology, 222: jeb 206045.

Williams Sandile E J, Wootton P, Mason Helen S, et al, 2004. Hemoxygenase-2 is an oxygen sensor for a calcium-sensitive potassium channel [J]. Science, 306 (5704): 2093-2097.

Winberg S, Olsén H, Höglund E, et al, 2011. Behavioural responses to hypoxia provide a non-invasive method for distinguishing between stress coping styles in fish [J]. Applied Animal Behaviour Science, 132 (3): 211-216.

Wiseman S, Osachoff H, Bassett E, et al, 2007. Gene expression pattern in the liver during recovery from an acute stressor in rainbow trout [J]. Comparative Biochemistry and Physiology Part D: Genomics and Proteomics, 2 (3): 234-244.

Wu C B, Zheng G D, Zhao X Y, et al, 2020. Hypoxia tolerance in a selectively bred F_4 population of blunt snout bream (*Megalobrama amblycephala*) under hypoxic stress [J]. Aquaculture, 518: 734484.

Wu Z, You F, Wen A, et al, 2016. Physiological and morphological effects of severe hypoxia, hypoxia and hyperoxia in juvenile turbot (*Scophthalmus maximus* L.) [J]. Aquaculture Research, 47 (1): 219-227.

Xiao H, Wang J, Yan W, et al, 2018. GLUT1 regulates cell glycolysis and proliferation in prostate cancer [J]. The Prostate, 78 (2): 86-94.

Xiao W, 2015. The hypoxia signaling pathway and hypoxic adaptation in fishes [J]. Science China Life Sciences, 58 (2): 148-155.

Xie J, He X, Fang H, et al, 2020. Identification of heme oxygenase-1 from golden pompano (*Trachinotus ovatus*) and response of Nrf2/HO-1 signaling pathway to copper-induced oxidative stress [J]. Chemosphere, 253: 126654.

Xue S, Lin J, Han Y, et al, 2021. Ammonia stress-induced apoptosis by p53-BAX/BCL-2 signal pathway in hepatopancreas of common carp (*Cyprinus carpio*) [J]. Aquaculture International, 29 (4): 1895-1907.

Xue Z, Zhang Y, Lin M, et al, 2017. Effects of habitat fragmentation on the population genetic diversity of the Amur minnow (*Phoxinus lagowskii*) [J]. Mitochondrial DNA Part B, 2 (1): 331-336.

Yahagi N, Shimano H, Matsuzaka T, et al, 2004. P53 involvement in the pathogenesis of fatty liver disease [J]. Journal of Biological Chemistry, 279 (20): 20571-20575.

Yang A, Kaghad M, Wang Y, et al, 1998. P63, a p53 homolog at 3q27-29, encodes multiple products with transactivating, death-inducing, and dominant-negative activities [J]. Molecular Cell, 2 (3): 305-316.

Yang H, Cao Z-D, Fu S-J, 2013. The effects of diel-cycling hypoxia acclimation on the hypoxia tolerance, swimming capacity and growth performance of southern catfish (*Silurus meridionalis*) [J]. Comparative biochemistry and physiology Part A: Molecular & integrative physiology, 165.

Yang S, Yan T, Wu H, et al, 2017. Acute hypoxic stress: Effect on blood parameters, antioxidant enzymes, and expression of HIF-1alpha and GLUT-1 genes in largemouth bass (*Micropterus salmoides*) [J]. Fish & Shellfish Immunology, 67.

Yang Y, Wang Z, Wang J, et al, 2021. Histopathological, hematological, and biochemical changes in high-latitude fish *Phoxinus lagowskii* exposed to hypoxia [J]. Fish Physiology and Biochemistry, 47 (4): 919-938.

Yi L, Ragsdale S W, 2007. Evidence that the heme regulatory motifs in heme oxygenase-2 serve as a thiol/disulfide redox switch regulating heme binding [J]. Journal of Biological Chemistry, 282 (29): 21056-21067.

Yin X, Liu Y, Zeb R, et al, 2020. The intergenerational toxic effects on offspring of medaka fish *Oryzias melastigma* from parental benzo [a] pyrene exposure via interference of the circadian rhythm [J]. Environmental Pollution, 267: 115437.

Yoo S, Yoo J, Kim H, et al, 2019. Neuregulin-1 protects neuronal cells against damage due to $CoCl_2$-induced hypoxia by suppressing hypoxia-inducible factor-1α and p53 in SH-SY5Y cells [J]. International neurourology journal, 23 (Suppl 2): S111-S118.

Zauner A, Daugherty W, Bullock M, et al, 2002. Brain oxygenation and energy metabolism: part I—biological function and pathophysiology [J]. Neurosurgery, 51: 289-301.

Zhang B, Chen N, Huang C, et al, 2016. Molecular response and association analysis of *Megalobrama amblycephala* FIH-1 with hypoxia [J]. Molecular Genetics and Genomics, 291 (4): 1615-1624.

Zhang E, Liu Y, Dentin R, et al, 2010. Cryptochrome mediates circadian regulation of camp signaling and hepatic gluconeogenesis [J]. Nature Medicine, 16: 1152-1156.

Zhang G, Zhao C, Wang Q, et al, 2017. Identification of HIF-1 signaling pathway in *Pelteobagrus vachelli* using RNA-Seq: effects of acute hypoxia and reoxygenation on oxygen sensors, respiratory metabolism, and hematology indices [J]. Journal of Comparative Physiology B, 187 (7): 931-943.

Zhang P, Lu L, Yao Q, et al, 2012. Molecular, functional, and gene expression analysis of zebrafish hypoxia-inducible factor-3 [J]. AJP Regulatory Integrative and Comparative Physiology, 303: R1165-1174.

Zhang P, Yao Q, Lu L, et al, 2014. Hypoxia-inducible factor 3 is an oxygen-dependent transcription activator and regulates a distinct transcriptional response to hypoxia [J]. Cell Reports, 6 (6): 1110-1121.

Zhang W, Xia S, Zhu J, et al, 2019. Growth performance, physiological response and histology changes of juvenile blunt snout bream, *Megalobrama amblycephala* exposed to chronic ammonia [J]. Aquaculture, 506: 424-436.

Zhang X, Sun Y, Chen J, et al, 2017. Gene duplication, conservation and divergence of Heme oxygenase 2 genes in blunt snout bream (*Megalobrama amblycephala*) and their responses to hypoxia [J]. Gene, 610: 133-139.

Zhang Y, Cao X, Gao J, 2019. Cloning of fatty acid-binding protein 2 (fabp2) in loach (*Misgurnus anguillicaudatus*) and its expression in response to dietary oxidized fish oil [J]. Comparative Biochemistry and Physiology Part B: Biochemistry and Molecular Biology, 229: 26-33.

Zhao Y, Jiang X, Kong X, et al, 2015. Effects of hypoxia on lysozyme activity and antioxidant defences in the kidney and spleen of *Carassius auratus* [J]. Aquaculture Research, 48.

Zhaparov B, Mirrakhimov M M, 1976. Ultrastructure of myocardial cells of yaks permanently living at high altitudes [J]. Bulletin of Experimental Biology and Medicine, 81 (6): 906-908.

Zhu C D, Wang Z H, Yan B, 2013. Strategies for hypoxia adaptation in fish species: a review [J]. Journal of Comparative Physiology B, 183 (8): 1005-1013.

Ziech D, Franco R, Pappa A, et al, 2011. Reactive Oxygen Species (ROS) —Induced genetic and epigenetic alterations in human carcinogenesis [J]. Mutation Research/Fundamental and Molecular Mechanisms of Mutagenesis, 711 (1): 167-173.

图书在版编目（CIP）数据

我国土著鱼类洛氏鲅的低氧耐受生理及其分子机制研究/母伟杰著.—北京：中国农业出版社，2022.8
ISBN 978-7-109-29742-5

Ⅰ.①我… Ⅱ.①母… Ⅲ.①鲤形目－淡水养殖－抗氧化作用（运动生物化学）－研究 Ⅳ.①S965.199

中国版本图书馆 CIP 数据核字（2022）第 129780 号

WOGUO TUZHU YULEI LUOSHIGUI DE DIYANG
NAISHOU SHENGLI JIQI FENZI JIZHI YANJIU

中国农业出版社出版
地址：北京市朝阳区麦子店街 18 号楼
邮编：100125
责任编辑：肖 邦 王金环
版式设计：杜 然 责任校对：沙凯霖
印刷：北京通州皇家印刷厂
版次：2022 年 8 月第 1 版
印次：2022 年 8 月北京第 1 次印刷
发行：新华书店北京发行所
开本：700mm×1000mm 1/16
印张：7.75 插页：6
字数：140 千字
定价：50.00 元

版权所有·侵权必究
凡购买本社图书，如有印装质量问题，我社负责调换。
服务电话：010-59195115 010-59194918

彩图1　洛氏鲅

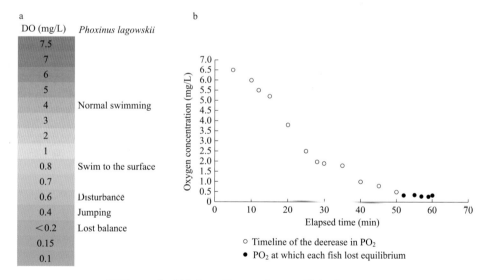

彩图2　洛氏鲅在不同 DO 水平下的行为和 LOE

a. 低氧胁迫期间洛氏鲅的行为　b. 洛氏鲅在氧气水平下降时的 LOE

DO 溶解氧　normal swimming 正常游泳　swim to surface 游至表面

disturbance 剧烈游动　jumping 跳跃　lost balance 失去平衡

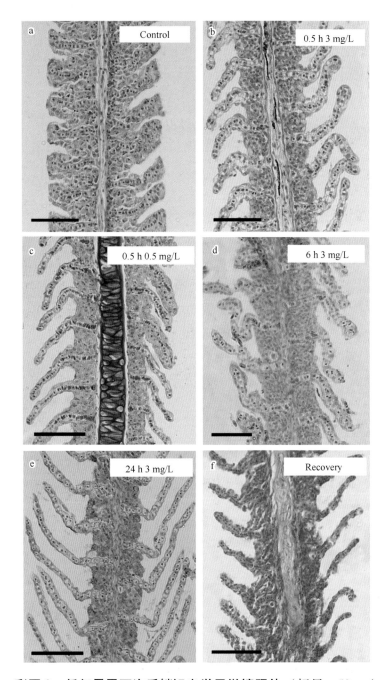

彩图 3 低氧暴露下洛氏鱥鳃光学显微镜照片（标尺 = 50μm）
a. 对照组　b. 0.5h 3mg/L 低氧处理　c. 0.5h 0.5mg/L 低氧处理
d. 6h 3mg/L 低氧处理　e. 24h 3mg/L 低氧处理　f. 恢复组
注：蓝色箭头代表空泡，黑色箭头代表肝细胞。

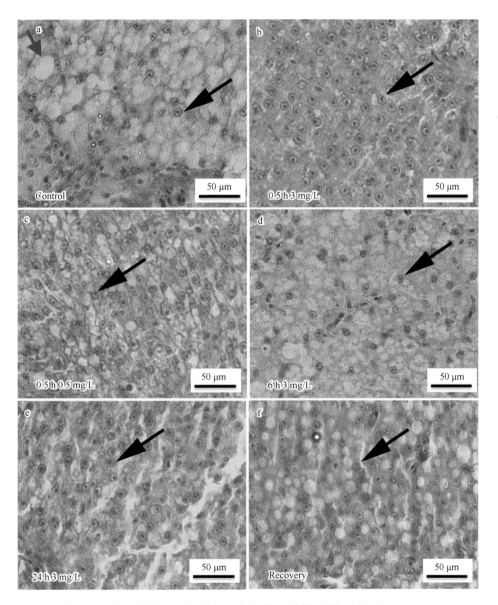

彩图 4　低氧暴露下洛氏鱥肝脏光学显微镜照片（标尺 = 50μm）
a. 对照组　b. 0.5 h 3mg/L 低氧处理　c. 0.5 h 0.5mg/L 低氧处理
d. 6 h 3mg/L 低氧处理　e. 24 h 3mg/L 低氧处理　f. 恢复组

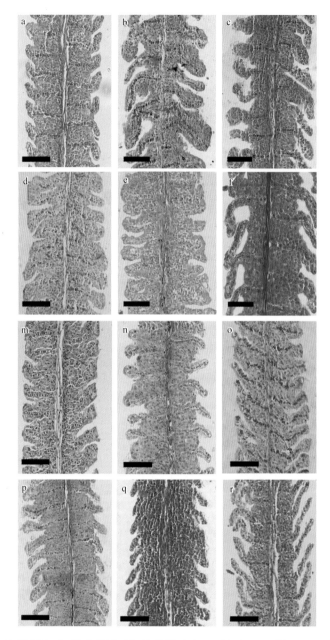

彩图 5 持续低氧和昼夜低氧处理后第 2 天、第 4 天、第 6 天、第 8 天和第 10 天洛氏鱥鳃的光学显微镜照片（标尺 =50μm）

a、m. 正常氧气　b. 持续低氧 2d　c. 持续低氧 4d　d. 持续低氧 6d
e. 持续低氧 8d　f. 持续低氧 10d　n. 循环低氧 2d　o. 循环低氧 4d
p. 循环低氧 6d　q. 循环低氧 8d　r. 循环低氧 10d

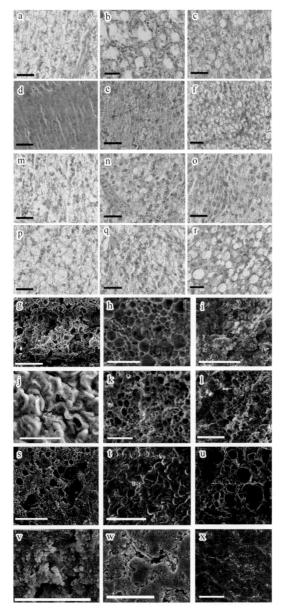

彩图6 持续低氧和昼夜低氧处理后第2天、第4天、第6天、第8天和第10天洛氏鱥肝脏的光学显微镜及扫描电子显微照片

(黑色标尺=50μm,白色标尺=100μm)

a、g、m、s. 正常氧气　b、h. 持续低氧2d　c、i. 持续低氧4d
d、j. 持续低氧6d　e、k. 持续低氧8d　f、l. 持续低氧10d　n、t. 循环低氧2d
o、u. 循环低氧4d　p、v. 循环低氧6d　q、w. 循环低氧8d　r、x. 循环低氧10d

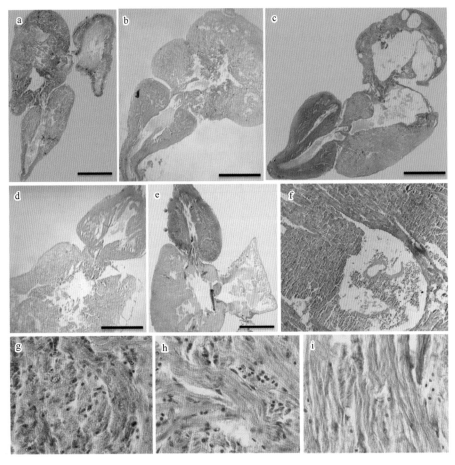

彩图 7　心脏经低氧处理后的 H&E 染色

a. 对照组心脏整体　b. 持续低氧 21d 心脏整体　c. 持续低氧 28d 心脏整体
d. 昼夜低氧 21d 心脏整体　e. 昼夜低氧 28d 心脏整体
f. 昼夜低氧处理 28d H&E 染色　g. 对照组心肌细胞核
h. 持续低氧 28d 心肌细胞核　i. 昼夜低氧 28d 心肌细胞核

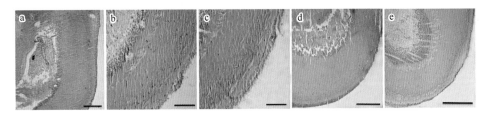

彩图 8　脑经持续低氧后的 H&E 染色

a. 对照　b. 持续低氧 7d　c. 持续低氧 14d　d. 持续低氧 21d　e. 持续低氧 28d

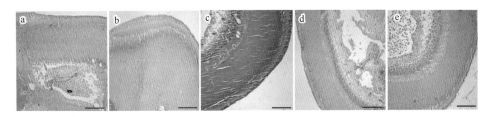

彩图 9　脑经昼夜低氧后的 H&E 染色
a. 对照　b. 持续低氧 7d　c. 持续低氧 14d　d. 持续低氧 21d　e. 持续低氧 28d

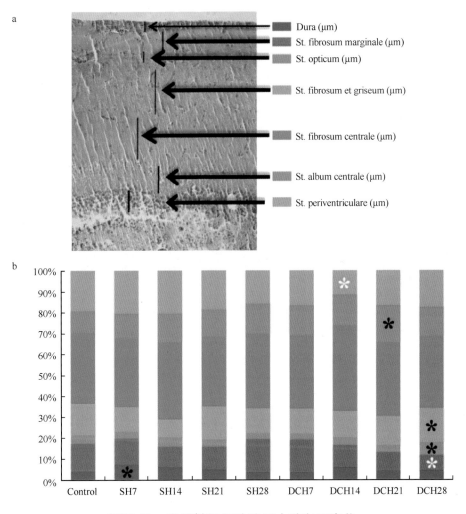

彩图 10　洛氏鲅低氧胁迫后中脑各层变化
a. 中脑分层示例　b. 中脑各层变化
注：＊代表各组之间存在显著差异（$P<0.05$）。

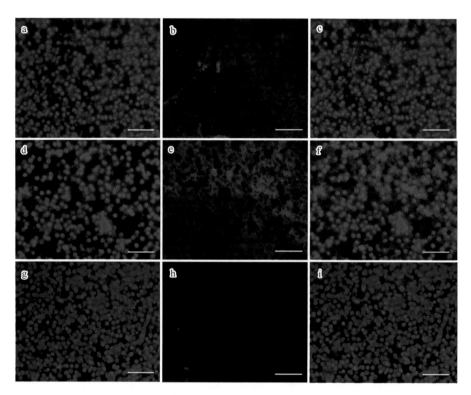

彩图11　荧光原位杂交检测肝脏中 *HIF-1α* mRNA 的表达（标尺 =50μm）

a、b、c. 带探针的对照组　d、e、f. 带探针的处理组

g、h、i. 无探针的阴性对照组

注：a、d、g 表示用 DAPI 将细胞核染成蓝色，c、f 表示 HIF-1α 信号和 DAPI 染色的合并显示。

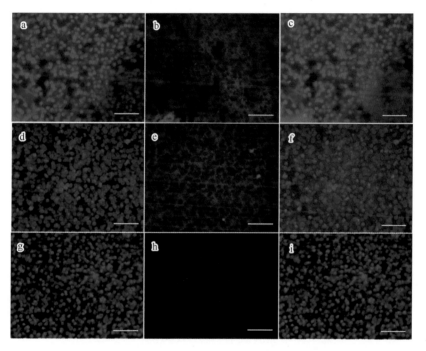

彩图 12　荧光原位杂交检测肝脏中 *HIF-2α* mRNA 的表达（标尺 =50μm）

a、b、c. 带探针的对照组　d、e、f. 带探针的处理组

g、h、i. 无探针的阴性对照组

注：a、d、g 表示用 DAPI 将细胞核染成蓝色，c、f 表示 HIF-2α 信号和 DAPI 染色的合并显示。

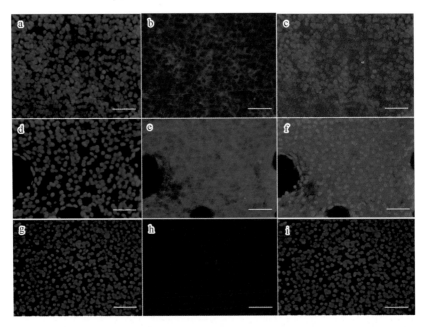

彩图 13　荧光原位杂交检测肝脏中 *HIF-3α* mRNA 的表达（标尺 =50μm）

a、b、c. 带探针的对照组　d、e、f. 带探针的处理组
g、h、i. 无探针的阴性对照组

注：a、d、g 表示用 DAPI 将细胞核染成蓝色，c、f 表示 HIF-3α 信号和 DAPI 染色的合并显示。

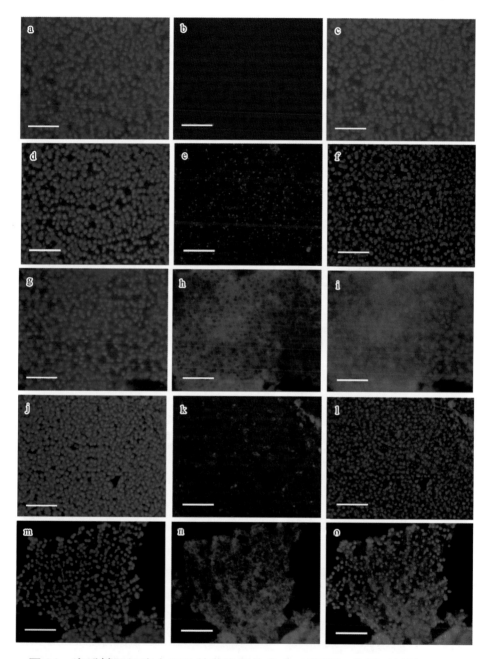

图 14 洛氏鲅肝细胞中 P53 的荧光原位杂交（FISH）分析（标尺 =50μm）

a、b、c. 阴性对照组　d、e、f. 常氧对照组　g、h、i. 24h 短期低氧组
j、k、l. 2d 持续低氧处理组　m、n、o. 2d 的昼夜循环低氧组

注：a、d、g、j、m 表示 DAPI 染色的蓝色荧光，b、e、h、k、n 表示探针染色的红色免疫荧光，c、f、i、l、o 分别为前两组的重叠。

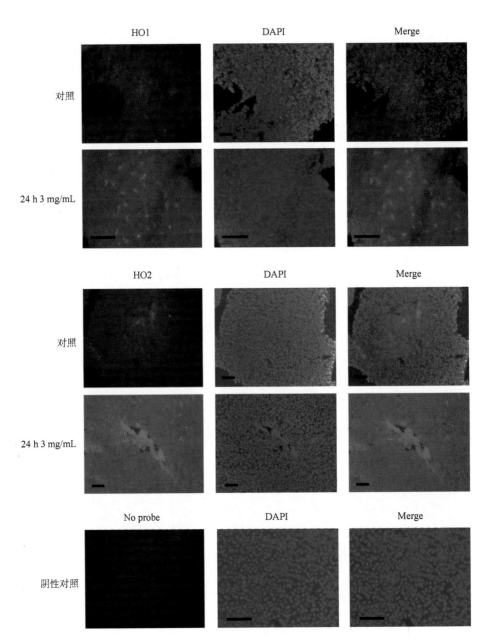

彩图 15　洛氏鲅 *HO-1* 基因和 *HO-2* 基因的原位杂交（标尺 =50μm）